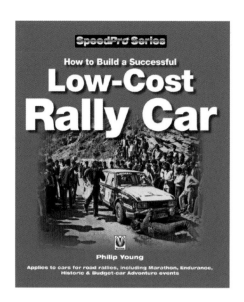

Other great books from Veloce –

Speedpro Series
4-cylinder Engine – How To Blueprint & Build A Short Block For High Performance (Hammill)
Alfa Romeo DOHC High-performance Manual (Kartalamakis)
Alfa Romeo V6 Engine High-performance Manual (Kartalamakis)
BMC 998cc A-series Engine – How To Power Tune (Hammill)
1275cc A-series High-performance Manual (Hammill)
Camshafts – How To Choose & Time Them For Maximum Power (Hammill)
Competition Car Datalogging Manual, The (Templeman)
Cylinder Heads – How To Build, Modify & Power Tune Updated & Revised Edition (Burgess & Gollan)
Distributor-type Ignition Systems – How To Build & Power Tune New 3rd Edition (Hammill)
Fast Road Car – How To Plan And Build Revised & Updated Colour New Edition (Stapleton)
Ford SOHC 'Pinto' & Sierra Cosworth DOHC Engines – How To Power Tune Updated & Enlarged Edition (Hammill)
Ford V8 – How To Power Tune Small Block Engines (Hammill)
Harley-Davidson Evolution Engines – How To Build & Power Tune (Hammill)
Holley Carburetors – How To Build & Power Tune Revised & Updated Edition (Hammill)
Jaguar XK Engines – How To Power Tune Revised & Updated Colour Edition (Hammill)
MG Midget & Austin-Healey Sprite – How To Power Tune New 3rd Edition (Stapleton)
MGB 4-cylinder Engine – How To Power Tune (Burgess)
MGB V8 Power – How To Give Your, Third Colour Edition (Williams)
MGB, MGC & MGB V8 – How To Improve New 2nd Edition (Williams)
Mini Engines – How To Power Tune On A Small Budget Colour Edition (Hammill)
Motorcycle-engined Racing Car – How To Build (Pashley)
Motorsport – Getting Started in (Collins)
Nitrous Oxide High-performance Manual, The (Langfield)
Rover V8 Engines – How To Power Tune (Hammill)
Sportscar & Kitcar Suspension & Brakes – How To Build & Modify Revised 3rd Edition (Hammill)
SU Carburettor High-performance Manual (Hammill)
Successful Low-Cost Rally Car, How to Build a (Young)
Suzuki 4x4 – How To Modify For Serious Off-road Action (Richardson)
Tiger Avon Sportscar – How To Build Your Own Updated & Revised 2nd Edition (Dudley)
TR2, 3 & TR4 – How To Improve (Williams)
TR5, 250 & TR6 – How To Improve (Williams)
TR7 & TR8 – How To Improve (Williams)
V8 Engine – How To Build A Short Block For High Performance (Hammill)
Volkswagen Beetle Suspension, Brakes & Chassis – How To Modify For High Performance (Hale)
Volkswagen Bus Suspension, Brakes & Chassis – How To Modify For High Performance (Hale)
Weber DCOE, & Dellorto DHLA Carburetors – How To Build & Power Tune 3rd Edition (Hammill)

Those Were The Days ... Series
Alpine Trials & Rallies 1910-1973 (Pfundner)
Austerity Motoring (Bobbitt)
Brighton National Speed Trials (Gardiner)
British Lorries Of The 1950s (Bobbitt)
British Touring Car Championship, The (Collins)
British Police Cars (Walker)
British Woodies (Peck)
Dune Buggy Phenomenon (Hale)
Dune Buggy Phenomenon Volume 2 (Hale)
Hot Rod & Stock Car Racing in Britain In The 1980s (Neil)
Last Real Austins, The, 1946-1959 (Peck)
MG's Abingdon Factory (Moylan)
Motor Racing At Brands Hatch In The Seventies (Parker)
Motor Racing At Brands Hatch In The Eighties (Parker)
Motor Racing At Crystal Palace (Collins)
Motor Racing At Goodwood In The Sixties (Gardiner)
Motor Racing At Nassau In The 1950s & 1960s (O'Neil)
Motor Racing At Oulton Park In The 1960s (Mcfadyen)
Motor Racing At Oulton Park In The 1970s (Mcfadyen)
Three Wheelers (Bobbitt)

Enthusiast's Restoration Manual Series
Citroën 2CV, How To Restore (Porter)
Classic Car Bodywork, How To Restore (Thaddeus)
Classic Car Electrics (Thaddeus)
Classic Cars, How To Paint (Thaddeus)
Reliant Regal, How To Restore (Payne)
Triumph TR2/3/3A, How To Restore (Williams)
Triumph TR4/4A, How To Restore (Williams)
Triumph TR5/250 & 6, How To Restore (Williams)
Triumph TR7/8, How To Restore (Williams)
Volkswagen Beetle, How To Restore (Tyler)
VW Bay Window Bus (Paxton)
Yamaha FS1-E, How To Restore (Watts)

Essential Buyer's Guide Series
Alfa GT (Booker)
Alfa Romeo Spider Giulia (Booker & Talbott)
BMW GS (Henshaw)
BSA Bantam (Henshaw)
BSA Twins (Henshaw)
Citroën 2CV (Paxton)
Citroën ID & DS (Heilig)
Fiat 500 & 600 (Bobbitt)
Jaguar E-type 3.8 & 4.2-litre (Crespin)
Jaguar E-type V12 5.3-litre (Crespin)
Jaguar XJ 1995-2003 (Crespin)
Jaguar/Daimler XJ6, XJ12 & Sovereign (Crespin)
Jaguar/Daimler XJ40 (Crespin)
Jaguar XJ-S (Crespin)
MGB & MGB GT (Williams)
Mercedes-Benz 280SL-560DSL Roadsters (Bass)
Mercedes-Benz 'Pagoda' 230SL, 250SL & 280SL Roadsters & Coupés (Bass)
Mini (Paxton)
Morris Minor & 1000 (Newell)
Porsche 928 (Hemmings)
Rolls-Royce Silver Shadow & Bentley T-Series (Bobbitt)
Subaru Impreza (Hobbs)
Triumph Bonneville (Henshaw)
Triumph TR6 (Williams)
VW Beetle (Cservenka & Copping)
VW Bus (Cservenka & Copping)
VW Golf GTI (Cservenka & Copping)

Auto-Graphics Series
Fiat-based Abarths (Sparrow)
Jaguar MKI & II Saloons (Sparrow)
Lambretta Li Series Scooters (Sparrow)

Rally Giants Series
Audi Quattro (Robson)
Austin Healey 100-6 & 3000 (Robson)
Fiat 131 Abarth (Robson)
Ford Escort MkI (Robson)
Ford Escort RS Cosworth & World Rally Car (Robson)
Ford Escort RS1800 (Robson)
Lancia Stratos (Robson)
Mini Cooper/Mini Cooper S (Robson)
Peugeot 205 T16 (Robson)
Subaru Impreza (Robson)

General
1½-litre GP Racing 1961-1965 (Whitelock)
AC Two-litre Saloons & Buckland Sportscars (Archibald)
Alfa Romeo Giulia Coupé GT & GTA (Tipler)
Alfa Romeo Montreal – The Essential Companion (Taylor)
Alfa Tipo 33 (McDonough & Collins)
Alpine & Renault – The Development Of The Revolutionary Turbo F1 Car 1968 to 1979 (Smith)
Anatomy Of The Works Minis (Moylan)
Armstrong-Siddeley (Smith)
Autodrome (Collins & Ireland)
Automotive A-Z, Lane's Dictionary Of Automotive Terms (Lane)
Automotive Mascots (Kay & Springate)
Bahamas Speed Weeks, The (O'Neil)
Bentley Continental, Corniche And Azure (Bennett)
Bentley MkVI, Rolls-Royce Silver Wraith, Dawn & Cloud/Bentley R & S-Series (Nutland)
BMC Competitions Department Secrets (Turner, Chambers Browning)
BMW 5-Series (Cranswick)
BMW Z-Cars (Taylor)
BMW Boxer Twins 1970-1995 Bible, The (Falloon)
Britains Farm Model Balers & Combines 1967 to 2007 (Pullen)
British 250cc Racing Motorcycles (Pereira)
British Cars, The Complete Catalogue Of, 1895-1975 (Culshaw & Horrobin)
BRM – A Mechanic's Tale (Salmon)
BRM V16 (Ludvigsen)
BSA Bantam Bible, The (Henshaw)
Bugatti Type 40 (Price)
Bugatti 46/50 Updated Edition (Price & Arbey)
Bugatti T44 & T49 (Price & Arbey)
Bugatti 57 2nd Edition (Price)
Caravans, The Illustrated History 1919-1959 (Jenkinson)
Caravans, The Illustrated History From 1960 (Jenkinson)
Carrera Panamericana, La (Tipler)
Chrysler 300 – America's Most Powerful Car 2nd Edition (Ackerson)
Chrysler PT Cruiser (Ackerson)
Citroën DS (Bobbitt)
Classic British Car Electrical Systems (Astley)
Cliff Allison – From The Fells To Ferrari (Gauld)
Cobra – The Real Thing! (Legate)
Cortina – Ford's Bestseller (Robson)
Coventry Climax Racing Engines (Hammill)
Daimler SP250 New Edition (Long)
Datsun Fairlady Roadster To 280ZX – The Z-Car Story (Long)
Diecast Toy Cars of the 1950s & 1960s (Ralston)
Dino – The V6 Ferrari (Long)
Dodge Challenger & Plymouth Barracuda (Grist)
Dodge Charger – Enduring Thunder (Ackerson)
Dodge Dynamite! (Grist)
Donington (Boddy)
Draw & Paint Cars – How To (Gardiner)
Drive On The Wild Side, A – 20 Extreme Driving Adventures From Around The World (Weaver)
Ducati 750 Bible, The (Falloon)
Ducati 860, 900 And Mille Bible, The (Falloon)
Dune Buggy, Building A – The Essential Manual (Shakespeare)
Dune Buggy Files (Hale)
Dune Buggy Handbook (Hale)
Edward Turner: The Man Behind The Motorcycles (Clew)
Fast Ladies – Female Racing Drivers 1888 to 1970 (Bouzanquet)
Fiat & Abarth 124 Spider & Coupé (Tipler)
Fiat & Abarth 500 & 600 2nd Edition (Bobbitt)
Fiats, Great Small (Ward)
Fine Art Of The Motorcycle Engine, The (Peirce)
Ford F100/F150 Pick-up 1948-1996 (Ackerson)
Ford F150 Pick-up 1997-2005 (Ackerson)
Ford GT – Then, And Now (Streather)
Ford GT40 (Legate)
Ford In Miniature (Olson)
Ford Model Y (Roberts)
Ford Thunderbird From 1954, The Book Of The (Long)
Forza Minardi! (Vigar)
Funky Mopeds (Skelton)
Gentleman Jack (Gauld)
GM In Miniature (Olson)
GT – The World's Best GT Cars 1953-73 (Dawson)
Hillclimbing & Sprinting – The Essential Manual (Short & Wilkinson)
Honda NSX (Long)
Jaguar, The Rise Of (Price)
Jaguar XJ-S (Long)
Jeep CJ (Ackerson)
Jeep Wrangler (Ackerson)
Karmann-Ghia Coupé & Convertible (Bobbitt)
Lamborghini Miura Bible, The (Sackey)
Lambretta Bible, The (Davies)
Lancia 037 (Collins)
Lancia Delta HF Integrale (Blaettel & Wagner)
Land Rover, The Half-ton Military (Cook)
Laverda Twins & Triples Bible 1968-1986 (Falloon)
Lea-Francis Story, The (Price)
Lexus Story, The (Long)
little book of smart, the (Jackson)
Lola – The Illustrated History (1957-1977) (Starkey)
Lola – All The Sports Racing & Single-seater Racing Cars 1978-1997 (Starkey)
Lola T70 – The Racing History & Individual Chassis Record 4th Edition (Starkey)
Lotus 49 (Oliver)
Marketingmobiles, The Wonderful Wacky World Of (Hale)
Mazda MX-5/Miata 1.6 Enthusiast's Workshop Manual (Grainger & Shoemark)
Mazda MX-5/Miata 1.8 Enthusiast's Workshop Manual (Grainger & Shoemark)
Mazda MX-5 Miata: The Book Of The World's Favourite Sportscar (Long)
Mazda MX-5 Miata Roadster (Long)
Maximum Mini (Booij)
MGA (Price Williams)
MGB & MGB GT– Expert Guide (Auto-doc Series) (Williams)
MGB Electrical Systems Updated & Revised Edition (Astley)
Micro Caravans (Jenkinson)
Micro Trucks (Mort)
Microcars At Large! (Quellin)
Mini Cooper – The Real Thing! (Tipler)
Mitsubishi Lancer Evo, The Road Car & WRC Story (Long)
Montlhéry, The Story Of The Paris Autodrome (Boddy)
Morgan Maverick (Lawrence)
Morris Minor, 60 Years On The Road (Newell)
Moto Guzzi Sport & Le Mans Bible, The (Falloon)
Motor Movies – The Posters! (Veysey)
Motor Racing – Reflections Of A Lost Era (Carter)
Motorcycle Apprentice (Cakebread)
Motorcycle Road & Racing Chassis Designs (Noakes)
Motorhomes, The Illustrated History (Jenkinson)
Motorsport In colour, 1950s (Wainwright)
Nissan 300ZX & 350Z – The Z-Car Story (Long)
Off-Road Giants! – Heroes of 1960s Motorcycle Sport (Westlake)
Pass The Theory And Practical Driving Tests (Gibson & Hoole)
Peking To Paris 2007 (Young)
Plastic Toy Cars Of The 1950s & 1960s (Ralston)
Pontiac Firebird (Cranswick)
Porsche Boxster (Long)
Porsche 356 (2nd Edition) (Long)
Porsche 908 (Födisch, Neßhöver, Roßbach, Schwarz & Roßbach)
Porsche 911 Carrera – The Last Of The Evolution (Corlett)
Porsche 911R, RS & RSR, 4th Edition (Starkey)
Porsche 911 – The Definitive History 1963-1971 (Long)
Porsche 911 – The Definitive History 1971-1977 (Long)
Porsche 911 – The Definitive History 1977-1987 (Long)
Porsche 911 – The Definitive History 1987-1997 (Long)
Porsche 911 – The Definitive History 1997-2004 (Long)
Porsche 911SC 'Super Carrera' – The Essential Companion (Streather)
Porsche 914 & 914-6: The Definitive History Of The Road & Competition Cars (Long)
Porsche 924 (Long)
Porsche 928 (Long)
Porsche 944 (Long)
Porsche 964, 993 & 996 Data Plate Code Breaker (Streather)
Porsche 993 'King Of Porsche' – The Essential Companion (Streather)
Porsche 996 'Supreme Porsche' – The Essential Companion (Streather)
Porsche Racing Cars – 1953 To 1975 (Long)
Porsche Racing Cars – 1976 On (Long)
Porsche – The Rally Story (Meredith)
Porsche: Three Generations Of Genius (Meredith)
RAC Rally Action! (Gardiner)
Rallye Sport Fords: The Inside Story (Moreton)
Redman, Jim – 6 Times World Motorcycle Champion: The Autobiography (Redman)
Rolls-Royce Silver Shadow/Bentley T Series Corniche & Camargue Revised & Enlarged Edition (Bobbitt)
Rolls-Royce Silver Spirit, Silver Spur & Bentley Mulsanne 2nd Edition (Bobbitt)
Russian Motor Vehicles (Kelly)
RX-7 – Mazda's Rotary Engine Sportscar (Updated & Revised New Edition) (Long)
Scooters & Microcars, The A-Z Of Popular (Dan)
Scooter Lifestyle (Grainger)
Singer Story: Cars, Commercial Vehicles, Bicycles & Motorcycle (Atkinson)
SM – Citroën's Maserati-engined Supercar (Long & Claverol)
Subaru Impreza: The Road Car And WRC Story (Long)
Supercar, How To Build your own (Thompson)
Taxi! The Story Of The 'London' Taxicab (Bobbitt)
Tinplate Toy Cars Of The 1950s & 1960s (Ralston)
Toyota Celica & Supra, The Book Of Toyota's Sports Coupés (Long)
Toyota MR2 Coupés & Spyders (Long)
Triumph Motorcycles & The Meriden Factory (Hancox)
Triumph Speed Twin & Thunderbird Bible (Woolridge)
Triumph Tiger Cub Bible (Estall)
Triumph Trophy Bible (Woolridge)
Triumph TR6 (Kimberley)
Unraced (Collins)
Velocette Motorcycles – MSS To Thruxton Updated & Revised (Burris)
Virgil Exner – Visioneer: The Official Biography Of Virgil M Exner Designer Extraordinaire (Grist)
Volkswagen Bus Book, The (Bobbitt)
Volkswagen Bus Or Van To Camper, How To Convert (Porter)
Volkswagens Of The World (Glen)
VW Beetle Cabriolet (Bobbitt)
VW Beetle – The Car Of The 20th Century (Copping)
VW Bus – 40 Years Of Splitties, Bays & Wedges (Copping)
VW Bus Book, The (Bobbitt)
VW Golf: Five Generations Of Fun (Copping & Cservenka)
VW – The Air-cooled Era (Copping)
VW T5 Camper Conversion Manual (Porter)
VW Campers (Copping)
Works Minis, The Last (Purves & Brenchley)
Works Rally Mechanic (Moylan)

www.veloce.co.uk

First published in December 2009 by Veloce Publishing Limited, 33 Trinity Street, Dorchester DT1 1TT, England. Fax 01305 268864/e-mail info@veloce.co.uk/web www.veloce.co.uk or www.velocebooks.com.
ISBN: 978-1-84584-208-6/UPC: 6-36847-04208-0

© Philip Young and Veloce Publishing 2009. All rights reserved. With the exception of quoting brief passages for the purpose of review, no part of this publication may be recorded, reproduced or transmitted by any means, including photocopying, without the written permission of Veloce Publishing Ltd. Throughout this book logos, model names and designations, etc, have been used for the purposes of identification, illustration and decoration. Such names are the property of the trademark holder as this is not an official publication.

Readers with ideas for automotive books, or books on other transport or related hobby subjects, are invited to write to the editorial director of Veloce Publishing at the above address.
British Library Cataloguing in Publication Data – A catalogue record for this book is available from the British Library. Typesetting, design and page make-up all by Veloce Publishing Ltd on Apple Mac.
Printed in India by Replika Press.

SpeedPro Series

How to Build a Successful
Low-Cost Rally Car

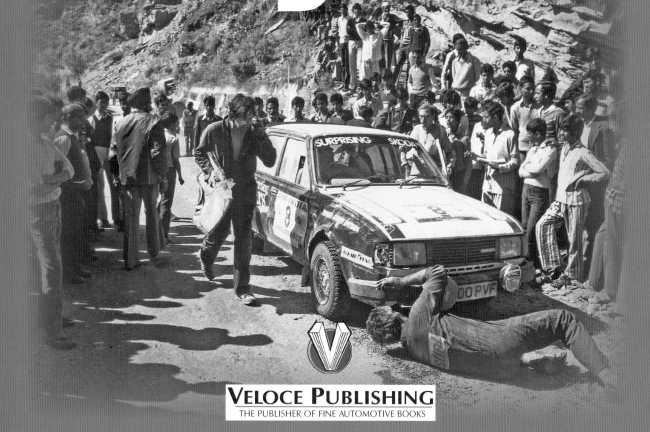

VELOCE PUBLISHING
THE PUBLISHER OF FINE AUTOMOTIVE BOOKS

Philip Young

Contents

Introduction 7
   To the start line 7

Acknowledgements 8

1. Austin Allegro project car 9
   The ten-day wonder. 9
   Job list 15
      Engine/mechanical 15
      Suspension 16
      Cooling 16
      Electrical 16
      Bodywork 16
      Protection 16
      Interior 16
      Safety 16
      Odds & sods 16
   Spare list 17
      Electrical 17
      Engine/transmission 17
      Brakes 17
      Fuel 17
      Suspension 17
      Tools 17
      Fluids 17
      Emergencies and odds & sods .. 17

2. Engines 18
   Oil options 23
   Carburettors 23

3. Radiators and cooling 25

4. Electrics 29
   Emergency electrical kit 31

5. Suspension 32

6. Brakes 35

7. Wheels and tyres 37

8. Bodyshells 40

9. Protection 45

10. Stickers 49
   Numbers 49
   Doorplates 49
   Crew names 50

**11. Navigation equipment
    & gadgets** 51

12. Safety 54

13. Keep it going 55
   The ten most common failings ... 55
   The most valued items in the
     tool bag 55
   The best roadside bodges to
     help you survive 55

14. Twenty bargain project cars ... 56
   Fiat Panda & Uno 56
   Nissan Micra 56
   Peugeot 205 57
   Rover 25 57
   Perodua Kalisa 58
   Proton Satria 58
   Hillman Hunter 59
   Austin 1800 59
   Suzuki 4X4 59
   Vauxhall Chevette 60
   Mercedes 60
   Peugeot 504 61
   Volvo GLT 61
   City Rover 61
   Morris Minor 62
   Ford Anglia 62

SPEEDPRO SERIES

Ford Ka 63
The Mini 63
Hillman Imp 64
Skoda Estelle 64

15. Driving tips 65
Sand 65
Snow and ice 66
River crossings 67
Brakes ... break it 68

16. Life on the road 69
Living with truckers 69
Meals on wheels 70
Toilet stops 71
Insurance 71
Calling International Rescue 71
Mobile phones 72

17. Border crossings 73
Paperwork required at borders 74
Dos and don'ts at borders 75
More useful tips 75

18. Your health 76
Taking care of number one 76

19. Sponsorship 79
A reality check 79

20. Event guide 81
British events 81
Historic events 82
   Minor earthquake rocks
   championship 83
The best of the British events ... 83
   For modern cars 83
   Welsh road rallying 84
   For Historic cars 85
   For Endurance cars 85
   For everybody 85
Foreign events 85
   Banger rallies 85
   Mongol Rally 86
   Plymouth Dakar 86

21. Useful contacts 88
Engine bits 88
Exhausts 88
Good oil 88
Suspension 88
Wheels 89
Transmission 89
Seatbelts 89
Seats 89
Spares 89
Brakes 89
Tyres 89
Rally car preparers 89
Stickers and numbers 89
Mail order parts and accessories . 89
Travel advice 89
Navigation 89
Clothing & camping 89
Map supplies 89
Books 89
Rally organisers 89
UK Endurance rallies 90
Governing bodies 90
Internet forums 90

Index 91

www.velocebooks.com / www.veloce.co.uk
All current books • New book news • Special offers • Gift vouchers

Veloce SpeedPro books -

 978-1-903706-92-3
 978-1-901295-26-9
 978-1-845840-23-5
 978-1-845840-19-8
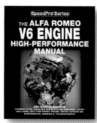 978-1-845840-21-1
 978-1-903706-59-6

 978-1-903706-76-3
 978-1-904788-78-2
 978-1-903706-78-7
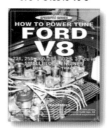 978-1-903706-72-5
 978-1-903706-94-7
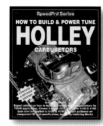 978-1-845840-06-8

 978-1-903706-91-6
 978-1-845840-05-1
 978-1-903706-77-0
 978-1-84584-187-4
 978-1-904788-93-5
 978-1-901295-73-3

 978-1-904788-84-3
 978-1-903706-70-1
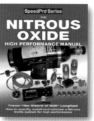 978-1-904788-89-8
 978-1-903706-17-6
 978-1-84584-207-9
 978-1-904788-91-1

 978-1-845840-73-0
 978-1-904788-22-5
 978-1-903706-80-0
 978-1-903706-68-8
 978-1-845840-45-7
 978-1-874105-70-1

 978-1-903706-14-5
 978-1-903706-99-2
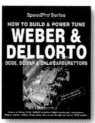 978-1-903706-75-6
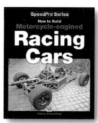 978-1-845841-23-2
 978-1-845841-62-1
 978-1-84584-208-6

Introduction

TO THE START LINE

This is not a book on how to get another ten horsepower out of the engine, or gain an extra tenth-of-a-second through a corner. It's about to how to get down from the bank under the trees, stop being just a spectator ... and join in. If you've always wanted to go rallying, but have been told it can only be done by raising an overdraft, this book is for you. If you think the world of rallying is all about high-technology, and high-expense, there are lots of fellow enthusiasts out there in a whole new 'alternative scene' who are demonstrating that there is another way.

Road events using the kind of car that can still do the school run or make the weekly drive to the supermarket are taking place the length and breadth of Britain every weekend of the year. One driver featured in this book is a district nurse who wins trophies at weekends in a 1-litre car that she uses all week on her nursing rounds.

Road rallying has been a stepping stone to greatness for many top navigators – one co-driver in today's World Rally Championship started off in night events on British lanes. British club road events have long been a breeding ground for developing and sharpening a cunning sense of ingenuity that can help a driver outwit a rival, but winning these events needs *two* people: a good driver and a good navigator. These rallies are not about who has the thickest cheque book, or the most mechanics backing up the most powerful car.

And there is now also a whole raft of overseas events, which until quite recently used to be the preserve of classic cars. This is something of a phenomenon. As the event guide at the back of this book reveals, there are dozens of budget car Adventure Rallies driving off to the four corners of the earth, from Kathmandu to Timbuktu. It's no exaggeration to say that there are now over 1000 entrants *every year* taking part in these events. The biggest rally in Britain is now the Mongol Rally – over 300 cars lined up to start the last one. Cheap old cars are being sought out to carry gap year students eager to see what the world is made of, and those who have managed to climb out of the rut of daily routine. A few days in a workshop could transform these ambitions, and the following pages attempt to explain how.

Hopefully, this book can also inspire you to take part, and with personal preparation advice of the kind not published elsewhere, will also help you survive the rigours of 'Adventure Road'.

Acknowledgements

This book could not have been produced without the assistance of Alan Smith, Heidi Winterbourne, Barbara Bradshaw, Simon Ayris (workshop photography at Rally Preparation Services, Witney), and the British Motor Heritage Museum, Gaydon. Additional technical support was provided by Jon Hill, Jim Gavin, Andy Inskip, Tony Fowkes, Den Green and Martin Kernahan.

Photography: Club Rally photography by M&H Photography and Brian Gilbert; Trials photos by Charles Wooding and Jonathan Toulmin; World Cup by Chris Bruce; Lombard by Gerard Brown; Plymouth-Dakar by James Cheadle; Mongol Rally and Africa Rally courtesy of The Adventurists Ltd.

Front cover: On the roof of the world! The author seen here on the 1985 Himalayan Rally arrives at a refuelling point, the works Skoda Estelle bearing all the signs of a long hard night – up to 3rd overall at this point. Chris Bruce, alongside, and Milosh from the Skoda factory under the front of the car, are about to carry out running repairs.

A good result was dashed on the last day when the gearbox mountings broke, and the engine and gearbox dropped onto the road, forcing the Skoda into retirement. The crew of Philip Young and Hywel Thomas now faced the tricky job of being towed 500 miles back down the mountains to Delhi. Searching for materials to get the engine and gearbox off the ground, and the driveshafts roughly level, Chris Bruce discovered an abandoned Scouts' hut on the edge of a nearby village, with an old piano inside. He cut the wires out of the back of the piano, laid them out on the road, and weaved them into a cat's cradle. Holes were drilled in the floor around the engine bay, and the wires poked up through and lashed onto the rollcage. A winch then lifted the engine and gearbox off the ground, suspended in the wire netting. The ends of the wires were tied, and with this lash-up the car was towed back to Delhi.

Make sure this doesn't happen to you – you might not find a handy Scouts' hut with an old piano. Check out the information on engine and gearbox mountings in this book.

Photo by Mike Johnson.

Chapter 1
Austin Allegro project car

THE TEN-DAY WONDER

This seemed such a totally hopeless lost cause. The idea was simple enough – drive an Austin Allegro two thirds of the way around the world, and hopefully pitch up for the last section of eight days of utter wilderness, where there are not only bad roads, but no roads at all, in a land so sparse you hardly see a soul from one day to the next. And if you do find someone, then you discover they don't speak English, have never seen an Allegro before, and certainly have no idea how Hydragas suspension works.

The Mongol Rally has taken off in a few short years and is now simply the biggest rally in Britain – no other event attracts so many competitors. At the time of writing, over 300 crews had lined up in London's Hyde Park. The first objective is to reach a checkpoint in a castle outside Prague, where everyone camps in the grounds for a party. Having done this you are on your own, and free to make your own way on the long drive to Mongolia.

Grand Prix of Mongolia ... cars start three abreast in Hyde Park, a Minor on pole position.

You have to reach the capital, Ulaan Baatar, to rank as a finisher. Many drop out. Reliability is the chief reason quoted – the car takes the blame. But we suspect thoroughness of personal crew preparations, relationships within

SPEEDPRO SERIES

Captain Blighty, with a highly suitable but totally unpatriotic car.

Austin Maestro – £500 to buy and prepare – departs Hyde Park.

"Which way for the Dover Road?" A 2CV makes its first stop for directions.

"Blimey ... you'll never make it, matey!" Our project Allegro in the Old Kent Road.

the car, a sudden desire to get home, money worries, food worries, health issues like simply not drinking enough, and failing to keep to a discipline of knocking off several hundred miles per day, all combine to sap the determination of reaching the end goal.

This, then, is the biggest, probably best-organised 'banger rally' around for low-cost budget cars. It's not just youngsters, students, and gap-year wanderers taking part; when we attended the official drivers' briefing at a café in London's Indian restaurant district of Brick Lane, we saw a few drivers who would qualify for a free bus-pass from their local council.

The chief requirement is that the vehicle should either be 'nutty' and of comedy value, or under 1000cc, as that makes it extra challenging. There is a general ethos that the car should be a rusty wreck, but as there are no rules or regulations other than the engine size, we can't see why you shouldn't enter a two year old Vauxhall Agila if you want to. The Mongolian government wants to ban leaving cars over ten years old in the country, which could change things.

So, that's the challenge. To make it even more tricky, the route splits up at Prague, with most taking a more boring trek across Russia. Some take a middle route through the Crimea to cross the Caspian Sea on a rickety, slow and uncomfortable flat-bottomed ferry, which creeps along at two miles an hour for fear of disturbing the shallow water, as the mud often clogs the propeller and stops all forward progress. Life on board is no picnic and the toilets are disgusting – welcome to the world's worst cruise liner.

For Dominic Spill, his sister Vicki, and friend Patrick Sumby, all in their mid-20s, taking an even more southerly direction via Turkey and crossing Iran appealed as the most challenging route of all. Not many take the Iran route, as it's the hottest time of the year (mid-July) and temperatures can climb to nearly 50C. You then have to cross 'The Stans' of Turkmenistan, Uzbekistan and Kazakhstan. Nearly 10,000 miles, with no back up, no doctors, no sweeper mechanics, no rally marshals to guide you when you're feeling low. You are on your lonesome, all the way.

AUSTIN ALLEGRO PROJECT CAR

Allegro Team: Patrick Sumby, Vicki and Dominic Spill.

Engine found on eBay, complete with special exhaust.

Plates welded to the front suspension area.

An extra plate in the side of the sill, plus two cans of builder's foam inside.

They could have chosen a better car! Their excuse for the Allegro? "We didn't know it was crap until we bought it – this car was, after all, made before our time." Then they saw it as adding to the challenge, particularly when someone in the Allegro Club told them it's not as bad as is made out ... but nobody has ever undertaken a journey as long as this one in an Allegro. "History is being made here," was their response to that.

They found 'Barney,' as they call their Allegro, on eBay, and it cost £250 – actually a fraction less, as the one-lady-owner felt sorry for them and gave a discount of £2 towards their bus fare. The car ran fine until it was pointed in the direct of Simon Ayris and the workshop of Rally Preparation Services. On an early morning run from Marlow in Buckinghamshire to Witney in Oxfordshire, the car conked out a few miles short of the destination. However, Dominic had only sat by the roadside for a minute or two when an AA van pulled up. "Are you a member?" "No, but I might be persuaded ..." "Then get on the end of this and see what you can do," says the AA man, handing Dominic a tow rope.

Barney arrived at the workshop, where it was discovered that the clutch had exploded in protest at the idea of going rallying. Simon promised to cap his time at exactly one week. Lots of jobs that could have been done were not even mentioned because of the time constraint. This is a cost thing here – it's all about keeping the costs to the minimum, while ensuring the car is prepped sufficiently to drive 40 days and 40 nights of hell on earth.

The engine was at the end of its life, and as it needed to be stripped down and given a clutch and second gear overhaul, it was suddenly decided to seek a completely reconditioned engine. Mini Sport at Padiham has a busy website and certainly looked an ideal solution, but then on eBay, and for less money, a complete engine and gearbox, with carburettor, alternator, and long centre branch manifold was on offer. A risk to rely on the complete unknown, but the risk was taken – the engine had been reconditioned by the Mini Centre of Heaton, Shropshire, and arrived on a courier's crate at Simon's workshop.

While the engine was out, Simon put some extra welding around the suspension areas of the engine bay. The sills were highly suspect, and rather than

SPEEDPRO SERIES

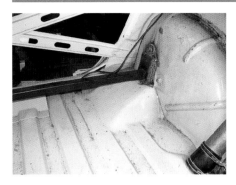

Crossbar welded over the suspension area across the width of the boot floor.

Relays protect the more powerful headlamp system.

Thick gloops of silicone bathroom sealer over the battery terminals and all electric joints. Note the Bacofoil heat deflector.

remove them and expose a house of horrors, the simple expedient of welding some sheets of metal along the inner edge and joining floor to sill with fresh metal, although a laborious job, added a lot of strength to a known weak area of the car. On top of that, two cans of builder's foam were injected into each sill, having first blocked some obvious holes underneath with underseal.

The suspension areas were reinforced, and a crossbar welded across the back of the boot to spread the load where the bump-stop rubbers take the worst of the suspension crashing metal against metal.

The electrics were tidied up, the headlights received relays, and all electrical connections were treated to blobs of bathroom silicone to keep out dust and water. With a new, diesel specification battery, the sides facing the engine were covered with kitchen foil as insulation.

Under the carpets, sheets of Thermawrap foil lagging were glued to the floor and tunnel with Evostick to fight heat-soak from the exhaust.

The exhaust itself received a morning's work as it was all new, routed differently to tuck up closer to the floor with a flexible joint. The tubing was heavy duty, thicker material and opened out to 1.75 with a straight through silencer, sourced from Jetex at Banbury, which specialises in bends and kinks for kit cars, vintage, classics and racing cars.

The suspension was left original – with a bigger budget, perhaps more could have been done with all-new Hydragas units, but time and money were strictly limited here. However, experiments with bigger bump-stop rubbers proved very useful. The thinking here was to limit some of the generous suspension travel

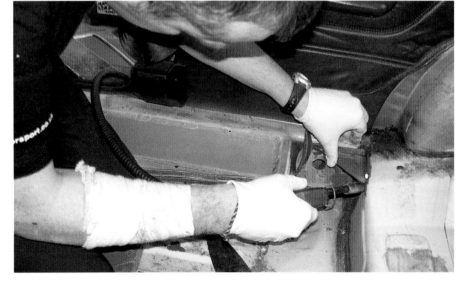

The triangle is a metal gusset, for strengthening the join between the sill and the under-seat beam. This helps prevent the floor flexing, and allows it to cope with extra weight.

Sheets of Thermawrap are stuck to the floor with Evostick, underneath the carpets.

AUSTIN ALLEGRO PROJECT CAR

Extra strong bump-stop rubber from Rally Design.

Reinforced metalwork around the rear suspension area, plus bigger bump-stops.

to save the Hydragas units. It was one of the cheaper mods to the car.

Standard steel wheels were retained, painted, and fitted with the tallest profile tyres that could be found that come with extra plies – in this case, Avon Avanza six-ply van tyres in 165-80-13, two sizes up on standard. The tyres alone gave an extra inch of ground clearance. Playing around with the Hydragas pressure gave another inch. Steel wheels are preferable – alloy ones chip and break very easily when hitting potholes, whereas steel buckles and bends, and can be beaten round again. Usually, a screw driver and a few taps with a hammer each evening sorts it out, but we have seen terrible cases – ultimately, though, if it's steel and bent, you are in with a chance. Alloy will just fracture (unless you have the famous Minilite, made of magnesium, but even these are mostly alloy these days).

A sump shield was found on the website of Mini Spares, which has a long list of bits and pieces and became a regular source of parts. This was the slatted steel version and at £20 was good value. It's not particularly robust, but is better than nothing; the attitude for the whole preparation is 'compromise.'

Avon Avanza van tyres give more ground clearance, and six plies help resist punctures.

A makeshift tow-bar across the front creates extra radiator protection. Taken from a scrapyard Delicia.

Some wire mesh between the slats was a final last minute mod. This is vital because often we see cases of a sump shield collecting fragments of grit or small stones, which lie on the upper faces of the sumpguard. The next time it's grounded, the enormous shock and force is enough simply to bang the stone through the bottom of the sump, and oil instantly pours out.

Two Metro seats were obtained for £20 each from a scrap yard, but not in a straightforward swap, so another

Two short scooter springs carried as spares for makeshift repairs, in case the Hydragas units give out.

SPEEDPRO SERIES

This Mini Spares sumpguard cost £20. It needed a sheet of polypropylene at the front for splash protection, to prevent water from shooting up to the distributor.

Are we sitting comfortably? Two MG Metro seats, £20 each from a scrapyard – big money for old seats. They took a morning to fit.

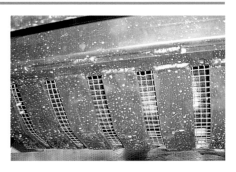

Some rabbit hutch wire netting, plus a sheet of rubber on top of the sumpguard, deflects stones and grit which could puncture the bottom of the sump pan.

morning was spent in the workshop making brackets. The interior was left fairly standard, but a word about cooling.

The trim that lines the C-pillars at the rear quarters of the car were removed to aid airflow through the gills, and this made a big difference. Someone in the back would be totally windswept if either of the front crew members had a side window down – now the back seat passenger would get the benefit of a more durable, constant breeze.

Drivers of works team Austin Healeys found that having an Austin A35 van vent in the rear of their hardtops dropped temperatures inside their fiery cockpits, and this little mod is copied today on modern rally cars with various roof vents. A hole in the roof protected by a cover adapted from a small 'poop scoop' shovel, obtained from a Wyevale garden centre and sprayed the body colour of the Allegro using aerosol paint, added a certain professional touch. (Cost: change out of a fiver.)

The engine bay can brew up fierce heat, particularly when idling or inching along in city traffic, so holes were cut at the back edge of the bonnet near the windscreen, principally to allow air to rise up and out here, and a bigger hole was cut as a vent further down, with a coffee-jar lid adapted into a 'power bulge'.

A piece of hardboard was cut to shape using a newspaper as a rough template to make a new boot floor, with the Thermawrap underneath to again hold back some heat from the exhaust.

The rear parcel shelf was treated to a thick layer of scrap carpet. Noise, heat, the very things that contribute to fatigue – if some cheap gizmo springs to mind

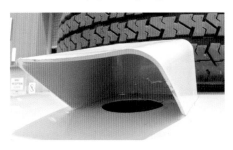

A Poop Scoop shovel from a garden centre, turned upside down and sprayed with aerosol paint, then glued to the roof – now an exit vent. Total cost ... £4.

Tear off the inside trim panel to the C-pillars to get more air flowing through the vents.

The vent has a chance to work without trim covering. It makes driving with the front windows open more pleasant.

K&N cotton air filter, with Bacofoil heat deflector glued to bulkhead with Evostick.

AUSTIN ALLEGRO PROJECT CAR

A power bulge in the bonnet covers the air vent. This was the lid of a coffee jar.

This white tubing is from a Hoover. It pipes cold air from the blanked-off side of the radiator grill to blast the hot area around the manifold, under the carb.

that will combat these problems, then it's probably worth a try.

Spare wheels were bolted to the roof, a plate having been put down the length of it first. Nuts that subsequently protruded through the headlining then got the plastic caps of Coke bottles glued over them, in a token gesture towards protection of someone's forehead while getting out of the back seat in a hurry.

The suspension is almost impossible to repair in the wild, and with three people travelling the car is overloaded before it starts – its Achilles' heel may prove to be the weight it had to lug up and down mountains and over the potholed scabby tarmac of some of the world's remotest deserts.

The build-sheet printed here may be useful as a guide to building another car for a project of this magnitude – many of the items could be applied to just about any car, not necessarily an Allegro. Another useful guide here is the spares list. This is interesting, as it's so long! All of the items are light by nature, and all of them seem to be necessary. When we sat down to compile it, we had in mind a couple of dozen items at most. It turned out to be three times longer.

Here is the biggest challenge of all – being totally disciplined in keeping the weight right down, personal luggage limited to just one soft bag per person, of the kind that goes inside an airline cabin. That should be the goal, with one more bag for spares, and a petrol can (one gallon of fuel, 5 litres, weighs 10lb). Water bottles are a must and can go in the rear door pockets. They'll be similar to the petrol can in quantity and weight, and that should be the limit.

Cost? The one week of labour turned into 10 days. The bits and pieces – £250 spent by the crew at an Allegro day – were a great help, but then the cost of the engine at over £1000 ruined the budget. It all came in at almost £5000, including the cost of the car. Twice what the team hoped for.

This was either the most expensive Allegro in the world, or a remarkably cheap car for driving over two-thirds of the earth, through some of the most stunningly beautiful and remote terrain, in the company of like-minded lunatics.

The Allegro completed the rally with no punctures. The only problem was a jammed driver's door.

JOB LIST
101 things to do to an Allegro:

Engine/mechanical
Recon 1275 engine with LCB manifold
Centre oil pick up pipe
Recon gearbox
New clutch
Dual 'see through' petrol filters
Engine breather pipe routed to top of inner wing
New brake master cylinder
All brakes inspected, lubricated, bled and fully adjusted
Resealed clutch master cylinder
New cotton air filter
Bespoke heavy duty exhaust system, flexible joint in middle section, larger single box
New oil pump
Loctite oil filter bolt
Stronger diff pin
Simplex timing chain

The simple stud on the roof is the same thread as wheel studs. The plate across the top holds extra wheels in place.

A bath plug chain to the oil cap means you'll never drop it in the dirt in the dark.

SPEEDPRO SERIES

Leaded conversion head
Halfords 20/50 running-in oil, change to Millers Transverse 20/50 oil before the start
New engine mountings, capped with chain to restrict movement

Suspension
Support bar across boot for rear suspension
Avon Avanza tyres 165/80/13
Castrol Molyslip grease on suspension joints and driveshafts
New front and rear wheel bearings assembled with Amsoil waterproof grease
Bump-stop rubbers 2in on front, MGB on back
Pumped up Hydragas units
Strengthened front and rear suspension mounting points
Rear suspension strap locating bolts modified
Overhauled front suspension

Cooling
New radiator – 3 core, export spec with increased capacity and flow
Flexible radiator mountings, Mini Cooper top, Mitsubishi Evo bottom
Antifreeze added to coolant to raise boiling point
Thermostat replaced by blanking sleeve to improve water circulation
Foil heat protection to fuel lines
Foil heat protection to bulkhead, foot wells, and floor
Holes in bonnet to extract air
Blanking panel to force air through radiator
Cold air ducting to carburettor
Bigger heat shield between manifold and carburettor/fuel lines
New cap for radiator expansion bottle

Electrical
New diesel-spec battery, foil wrapped, mounted on rubber mat
New copper core plug leads numbered and sequence numbered on engine block
New coil, electronic ignition, standard metro distributor
Auxiliary power socket
Relays added to headlight circuits
Halogen headlight bulbs

Bodywork
Steel plate to strengthen roof
Two wheels mounted on the roof
Rear roof air vent (Healey style)
Support bar across front of engine bay
Tow hooks front and rear
Resealed screen rubbers
Sun visor strip front and rear
Large mirror on passenger side
Trim removed from rear quarter panel vents
Bumpers, sills and wheels painted
Underseal to waterproof sills
Steel plate to strengthen floor and sills
Additional welding to seams in boot floor and wheelarches
Bar welded across boot between suspension mountings
Sills filled with expanding foam
Holes and gaps in bulkhead filled silicone sealant
Bostik around fuse box for airtight fitment into bulkhead

Protection
Additional steel tube front bumper nudge bar
Plastic splash guard to protect distributor
Mud flaps on all four wheels
Steel sumpguard with foam under sump
Silicone sealant on all electrical connections
Seal leads into distributor cap
Rubber boots over battery terminals, lined with silicone sealer
Additional welding to suspension mountings
Additional welding to front suspension in engine bay and to bulkhead
Additional welding around bump-stop area
Polypropylene fuel tank guard
Wire mesh in front of radiator
Steel conduit to protect Hydragas, brake and petrol pipes
Skid plate added to front of silencer box
Sponge padding between sumpshield and bottom of sump
Heat shield between fuel tank and exhaust
Conduit protection to brake lines in exposed areas

Interior
Reclining front seats (ex MG Metro)
Map and interior lights
Dashboard mounted cooling fan
Alarm clock on dashboard
Compass on dashboard
Map pockets on door panels and on back of front seats
Concealed waterproof container for documentation
Sound proofing and new boot floor
Carpeted rear parcel shelf
Coke bottle tops to cover roof bolts
Stick-on type interior light under bonnet
Coconut footwell mats (used under the wheels to aid traction)

Safety
Laminated windscreen
Rainex coated windscreen
Rear seat belt
Hand-held fire extinguisher

Odds & sods
Retention chain on fuel, oil and water filler caps
Headlamp tape to change to left hand dip beam
Wiring and piping tidied and secured with tie-wraps
New engine/chassis plate
New wiper blades
Air horns fitted
Spare ignition key taped inside rear bumper

AUSTIN ALLEGRO PROJECT CAR

SPARES LIST

Don't leave home without it ... the minimum spares kit for the Mongol Rally Allegro:

Electrical
Distributor cap and leads
Electronic distributor module
Complete electric fan unit
Fuses, relays – various
Bulbs – various
Connectors – assorted
Wire – assorted

Engine/transmission
Throttle cable
Throttle spring
Head gasket set
Clutch master and slave cylinder repair kit
Fan belt
Water pump
Exhaust joining pipe, clips and clamps

Brakes
Brake master cylinder repair kit

Fuel
Carburettor bits, jets, floats
Petrol pump
In-line fuel filter

Suspension
Motorcycle springs x 2, in case rear suspension collapses

Tools
Spanner set
Plug spanner
Adjustable spanner
Mole grips
Files, flat and round
Hacksaw and blades
Pliers, short and long
Stanley knife
Hammer
Screwdrivers, flat and Philips

Fluids
Brake fluid
Water
Petrol
Petrol syphon pump

Radweld
Redex to clear fuel lines
WD40
Petrol octane and lead booster additive
Funnel with filter

Emergencies and odds & sods
Temporary plastic windscreen or goggles
Windscreen sun shade
First aid kit
Fire extinguisher
Tow rope
Jump leads
Warning triangle
Gloves
Brolly
Torch and batteries
Penknife
Car manual
Tyre pressure gauge
Wet wipes
Bungy straps
Tie down straps
Jack and wood to support
Wheel brace
Spare wheel stud and nuts
Spare car keys
Wiper blade
De-ditching shovel
Puncture repair kit
Inner tube
Foot pump
Cable ties
Tank tape
Insulating tape
Plastic/rubber tubing
Jubilee clips, large and small
Instant gasket
Exhaust gum-gum
Hose bandage
Araldite
Wire, thick and thin (coat hanger)
Nuts, bolts, self tapping screws, split pins
Grommets, washers
Opal Fruits (now called Starbursts) to stop petrol tank leaks

The longest journey ever undertaken by an Allegro begins ...

Chapter 2
Engines

It might be stating the obvious, but the engine is the heart of a car. Bits that hang on the side of it, such as the exhaust, alternator, carburettor, fuel pump, filters, and distributor, give trouble on events from time to time and can, at the cost of losing time, be patched up and sorted, with the hope of being able to continue. Once the problem is inside, however, it demands open-heart surgery – serious.

You are not in a ten lap sprint around Brands Hatch; it's not a race where another 2bhp is going to make all the difference. However, most preparation workshops know only one approach – that of the club racer, looking for every extra ounce.

Reliability is what matters. Rebore it and it will run hotter, the pistons now much closer to the water jacket. Will that make for a better result? If it takes you outside the class limit, even by a few cc, you now have to keep a secret and hope nobody suggests popping something down a plughole to measure your engine. Put in a fancy crank to 'stroke' it and, yes, you can give it more torque, but there is always a downside. If the gearbox, or more likely the axle, is getting near its limit of tolerance as an abused road car, the manufacturer will never have dialled in the kind of reserves in the design to take the abuse of hard on-off, on-off acceleration and the reverse thrust of braking into corners that comes from charging down a timed rally section. The result is that within a few miles, something has to give.

Reliability first, performance second, has to be the approach here, the old adage being "to finish first, first you must finish." An engine problem is doom and gloom, and you don't have time to sort it. On longer events, an engine-out strip-down, even if spares are to hand, might just be possible with a view to getting back in the hunt again, but that rarely happens.

Cooling is a major source of problems, which is why we have devoted the longest chapter in the book to this.

The next big bugbear concerns exhausts.

It's almost true that you can get 'something for nothing' with an improved exhaust system; that is, a few simple and fairly inexpensive modifications will give you an improvement in performance.

Nissan Micra. Note the blue tube, left, and black tube, right, to push cold air to the carb, plus the motorcycle-style plug leads for waterproofing.

ENGINES

Neat work! Fuses clearly labelled, a Mini Cooper at the start of the Lombard.

Exhaust mounting bracket turned on its side for the Allegro, so the nuts and thread don't get damaged.

The downside, and there is always a downside to anything, is that you get more noise, which contributes to fatigue, and you could get more heat just where you don't want it.

The Allegro build in this book is a case in point. The reconditioned engine came with a free-flow, larger-bore exhaust manifold. It might contribute two or three extra horsepower, so it's a bonus. The downside is that the steel tubing lets out far, far more heat right under the carburettor than the original cast iron unit, and the standard manifold on an A+ engine is already quite a nicely designed, efficient manifold. Apart from the heat, the tubular manifold is not as quiet as the original cast iron one.

A common mistake is lagging the pipes with asbestos tape. It's a time consuming job to do well, so if you are asking a workshop to do it, don't baulk at the labour bill afterwards. You are now in trouble, as the incredible degrees of heat being forced down the pipe eventually burns through the metal walls of the piping. People just don't believe this, as they have never seen it occur, but run it day after day after day on a long event in hot weather and it will surely happen, as it did to Bob Fountain's Lagonda on the Peking to Paris. When he peeled off the layers of foil and asbestos, there were holes you could poke a thumb through – virtually nothing was left of the original manifold, and all he could do was limp on, sounding like a Lancaster bomber on take-off. He had yet to reach the halfway point of the 10,000 mile route. (And now had to find a workshop to make him a manifold – which he did, but the time loss ruined his promising placing in the classification.) Lagging the exhaust manifold, then, is a big 'No.'

The other issue concerning exhausts is the even more common failing of not mounting it with complete flexibility. It should be loose enough for the whole system to rock a little in your hand if you get hold of the end of the exhaust and shake it about. All the mountings should be flexible, and capable of withstanding some knocks. Don't have clamps with the nut and thread facing downwards, just because it's easy to do up when under a ramp – that is the first thing to dig into the dirt, hit a rock. Do you need these clamps, at all?

The best joint between pipes is the 'slipper joint' where one pipe slides deep inside another (at least three inches, more is better). You then tie the two pipes together, basically, with a figure of eight loop of wire, and here a wire coat hanger is brilliant. Two nuts on one end, on each side of the pipe, now, two nuts on the pipe you are joining to, brazed or mig-welded to each side (see drawing opposite), and you can now put in a loop of wire, and twist, and twist, so they are tight. The principle here is that if you belt the exhaust further back, like hitting a rock at 60mph with the silencer, the whole lot can take the punishment and the wire will come undone, dropping the exhaust system on the ground behind you, but it will be repairable and virtually undamaged.

The best flexible joint for exhausts ... nuts are welded to each side of the join.

SPEEDPRO SERIES

Coat hanger wire is now run through the nuts, in a figure of eight.

If a rock hits the silencer, this can spring apart. The exhaust will hit the ground with far less damage, and engine rock is absorbed thanks to supreme flexibility.

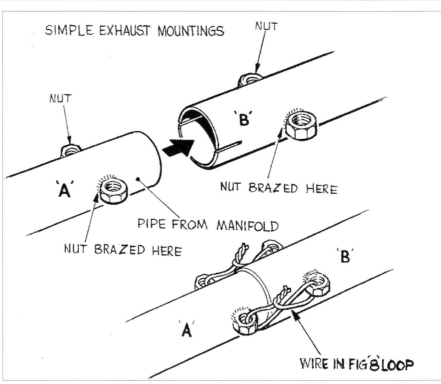

A master plan of the best exhaust system you can make – cost is one wire coat hanger and four nuts.

This Riley Pathfinder has the exhaust nicely tucked up out of the way – getting at the spare wheel could be a hassle ... nuts under the leaf springs look vulnerable.

A simple skid in front of the Allegro silencer box will help in desert crossings.

On my Paris Dakar Land Rover the exhaust was totally floppy, all the pipes joined together with wire. It was held up to the floor in a couple of places, but with rubber loops. It flexed, it banged against the floor often, but it never came undone, and it never leaked – we didn't gas ourselves. We had other problems, but exhaust issues were never on the nightly job lists. The final thing to say about exhausts is, the silencer. If it's free flow, if it's slightly more noisy, just remember you have got to put up with it hour after hour. Apart from heat, nothing is more tiring, more wearing, than that booming, drumming noise from the exhaust system.

The front of the silencer probably could do with a skid plate angled at around 45 degrees, to help it ride over a ridge coming out of a river, or glide over a brick in the road. Anything with right angles or sharp edges that face the direction of travel under the car should be skidded as a matter of course. It could also make a vital difference in sand, as it doesn't take much extra resistance from the bits that stick down to stop momentum, as you struggle to avoid finally coming to rest and getting totally stuck.

ENGINES

Stainless steel, or ordinary steel? If stainless is harder to weld, and the village blacksmith has never seen the stuff and doesn't speak English, then why spend the extra on it?

If you are rebuilding the engine, the questions will come up about skimming the cylinder head, flat-top pistons, and the like, which will raise compression ratios. You need to consider the fuel the engine is being asked to burn. If it's a great leap into the unknown – i.e. you are crossing the Sahara Desert to Cameroon, or some other far-flung destination – the chances of you completing the course without at least one tank of dodgy fuel are very remote.

Simon Ayris spent just 20 hours preparing a brand new City Rover 1.4 for two lads from Southampton University to drive to India, and they had only one mechanical problem – a sticking throttle cable which stuck in the flat-out position due to ridiculously high ambient temperatures under the bonnet, and one dose of poor petrol, which, surprisingly, happened not in Iran, or on reaching India, but in central Turkey.

Many modern cars have a shut-down device which can put an engine into limp-home-on-ultra-low-revs mode if the brain in the computerised engine management system detects a problem, like running on poor fuel. And poor fuel may not just mean low octane (as low as 70 is the norm in Mongolia, and 80 is pretty common in several African countries, and totally widespread in Russia). Poor fuel can be caused by locals lacing it with water. The answer to this is a really good fuel filter. The best is the Lucas truck or tractor item, with a button on the underside; tickle this and you can drop out the contents. Petrol floats on top of water, so it's a straightforward job to drop out the water regularly as it forms into large candle grease-like droplets and sinks to the bottom of the filter along with the dirt.

There are other filters – small items a couple of inches long made of plastic with a paper element inside. Put two or three into the fuel-lines. It's easy to undo the Jubilee clip round each end that pegs the fuel pipe to the ends of the filter. Nowhere near as idiot-proof (try doing this in the dark) as the Lucas jar, but better than nothing at all, and as they are small and cheap you can take several spares. So, this is one of the cheapest get-you-to-the-finish gadgets you can buy.

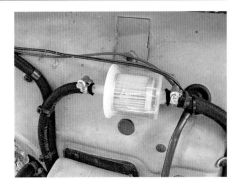

A simple cartridge device that goes in a petrol pipe is better than nothing, but not big enough. However, it costs just 40 pence.

Compression ratios were mentioned a moment ago. If you are racing, then you need every ounce of power. However, you are not racing, so why risk upping the compression ratio? Will ten horsepower make so much of a difference to your result? If the fuel is 70 octane then you need a compression ratio that is ideally 7:1, and if you know how to retard the ignition from time to time, then you might just get away with 8:1. No higher – that's the limit. Once you are up to 9:1 then you need modern fuels of the type here at home, with an octane ratio of 92 or more.

What is the downside? Two cars on the '97 Peking to Paris burned holes in their Heppolite pistons because their owners ignored this advice. An 1800 Landcrab was performing brilliantly across Tibet, its long wheelbase, long travel hydrolastic suspension was in its element, reminding us all of how Paddy Hopkirk so nearly won the London to Sydney Marathon in 1968. Then it burnt a hole in a piston. It had to limp on, with just three cylinders, all the way to the end in Paris – a wonder it lasted that long. The second car was a Morris Minor, which went off for a rebore overnight in Istanbul while a tired and worried crew went to bed – they woke up and found the local motor club had taken out the

Something from a Christmas tree? No, just all the metal filings and chaff that swim around in the oil at the bottom of the sump, stuck to the magnetic sump plug of the Allegro.

An old-fashioned petrol filter (left) is brilliant, if you can find one ...

engine, rebored just one cylinder, and fitted the only piston it had, a much bigger one from a Fiat, but it was now firing on four cylinders and it rode on all the way to Paris. Had the compression ratio been kept down to run poor local fuel, these problems could have been avoided. In other words, good results were chucked away because of the hunt for a few extra horses. Too high a compression ratio will see poor fuel burn valves and holes in pistons. Keep an ear open for 'pinking' – the distinctive chirping noise of an engine suffering from poor detonation.

It goes without saying, however, that if a cylinder head is coming off, and it's an older classic car, then you should take it to a specialist to have the valve seats upgraded to run unleaded fuel. Don't rely on plastic bottles of wonder additive. And talking of bottles of additive, do octane boosters work?

You would need a trailer to carry crates of the stuff on a long-distance epic. On the side of the bottle, it usually promises to boost octane levels by a factor of two, or even four. Wow! So you have just put in a tank load of fuel in the middle of Russia, you know its 80 octane (it says 90 on the pump – don't believe it, the hole in the ground is always the same stuff; they even write 95 on the pumps in Mongolia but the fuel is all 70 octane), so you pour in a bottle of octane booster. You have now boosted it by a maximum of four octane, but it won't make any difference that you will notice.

On the Himalayan Rally, we used Aldon octane booster to pour into engines like the TR8 and the Rover SD1, and the only thing it probably boosted was the driver's tired brain – engine strip-downs afterwards showed that just a few thousand hard miles had burned and chipped away at the edges of all the exhaust valves. So much for octane boosters.

Some garden chain-link wrapped around the engine mounting prevents upward snatch and the mounting snapping. Ideally, the chain should be tighter than this.

Other engine issues? We see broken engine mountings all the time. It was one of the factors that put me out of the running on the Paris Dakar, and my Land Rover had the uprated, heavy duty specials that were supposed to be unbreakable. You need to restrict the engine movement. An engine rocking forwards a touch will ensure the fan skims the radiator. Worse is to come – it could end up just resting on the chassis rails, as was the case with my drive in an ex-works Healey. Broken mountings caused the gearbox to drag on the ground of my works-built Skoda Estelle. It's a common problem. Capping upward movement is vital in a modern hatchback, as all cars today have engines located with mountings that enable a whole engine to break free easily in an accident. On any bumpy road you can ruin the engine mountings, and the first sign of trouble is usually that the exhaust is being whipped around and snapping the exhaust joints. There are more exhaust problems caused on long-distance events by worn engine mountings than anything else.

Minis and 1800 engines have a top stay with rubber at each end which helps to absorb the vibration, and some engines even have little shock absorbers. Capping the rocking of a heavy engine should be near the top of any rally preparation job list.

Fuel pumps are a regular complaint. The good old SU pump of Jaguars (and the smaller version of Morris Minor and Sprites) could be problematic when the points stick together, but you can take them apart and clean them out. The solid-state electronic pumps are supposed to be high-tech, and therefore vastly more trouble-free. I've used both, and had problems with both. The SU can be coaxed back into ticking again

ENGINES

(and it's rather assuring to be able to hear something ticking – at least that tells you it's pumping). On my Himalayan Rally Morris Minor, the Rev. Rupert Jones would belt the pump with his shoe. It would then make up for going on strike with a drum-beat of mad ticking until fuel supplies restored things to normal. It was part and parcel of the quirks of the car.

Facet pumps let me down totally on a Himalayan Rally with a turbocharged Range Rover, as they couldn't stand the high under-bonnet temperatures. Jan Odor at Janspeed said the pumps should be lagged with asbestos foil. That advice was not taken up, just missed off the job list. It might have taken a mere ten minutes, but it didn't get done – and the pumps packed up when we were third overall and going well. A frustrating, limping session all the way back down the hills to Delhi was a sad end to that effort. So, are solid-state pumps better? Maybe, maybe not – one thing is sure, when they go wrong, they have to be thrown away, they cannot be taken apart and fiddled with like the older technology – so pack a spare.

The other kind of pump is the mechanical pump attached to the engine block. Older Ford engines had these, with a handy little lever underneath that you could push up and down to crank up some fuel to prime the pump, all dead simple and usually trouble free. You need to ensure they get fuel well filtered, are kept cool (lag with asbestos), and you have a spare, as fuel pumps weigh little. Carry a spare and Murphy's Law says you won't need it. Risk going without, and Murphy's Law says you will be stopping somewhere remote, in the middle of the night, wondering why it won't restart.

Many cars have the air-intake mounted low down as it reduces noise inside the car, but it can then act as a scoop during a river crossing, sucking water into the engine with dire results.

OIL OPTIONS

All engine oil is brilliant stuff these days. Gone are the times when cars needed to go into a garage for a change to winter oil (thinner) and run thicker oil for the summer (Castrol XL 30 and Castrol XL 40 were designed for this).

Multi-grades came along as a result of the rush of technology brought about by World War II, Duckhams 20-50 being the first – well, Spitfire Merlin engines had to cope with -40 degrees in Russia and +40 degrees in Tunisia. We got the benefit in the 1950s when oil improved markedly, and just as well, as by the time the first motorway opened in 1960, and the Mini arrived with gears in the sump, oil simply had to be better.

I used to use Amsoil (stands for American Oil), which was the very first oil company to invent synthetic oil; expensive, but it worked brilliantly.

Today, all oil is brilliant and can cope with whatever you are asking of it, including running vastly longer distances between changes.

Penrite and Millers might not be household names, but they produce a range of oils, including special brews for classic and vintage cars that don't like modern thin oils full of detergent. Check out their websites from the list of useful contacts at the back of this book. Millers produces oils for engines with gearboxes in the sump, rated at E-P (extra pressure), which is demanded of gearbox lubrication.

The biggest no-no is putting modern thin oil into a seriously older engine design with large oil-ways, as those engines were designed for thick stuff. You can wear away the end of the distributor skew-gears, which work on the camshaft, which affects your engine timing, so you bang and pop along with

Get Amsoil grease – it's the best.

a seriously retarded engine – that's one downside of running the wrong oil.

Castrol does a non-multigrade 50 and 60-grade, as do Penrite and Millers, and this is what you should start out with if you are running an older engine. Otherwise, a sump full of 20-50 will suit a classic or worn engine. Modern engines such as a hatchback might be better off with a diesel oil, such as Castrol Magnatec, in 20-40 grade, available worldwide.

Oil is a personal thing; a bit like tyres, you will find everyone has their pet likes and firm opinions on what is best. However, the rule when going long-distance always has to be 'choose something good to start with, and something that will mix well with whatever can be found en-route, even if it's local truck oil.' Lugging around your own oil is just overloading the cargo.

CARBURETTORS

A change of carburettor could be a way of getting something for nothing, in that

SPEEDPRO SERIES

No heat-shield behind the three carbs of this Riley. Simply wrapping some pieces of asbestos cloth is better than nothing when it comes to preventing fuel vaporisation.

New K&N cotton filter. An old-fashioned oil-bath air cleaner might be better.

it might, if set up well, produce better breathing and therefore more power. This is unlikely to happen, however, if the camshaft is not lifting the valves a touch more, and if the valves and cylinder head have not been improved – a carb change on its own is likely to produce nothing more than a bigger fuel bill.

The key thing to consider here is the air filter. Modern K&N type cotton air filters are certainly very good, very efficient, and if set up properly, breath so well you can gain the advantage of one or two more horsepower. They are light, so take a couple of spares as they will clog up in no time in dusty conditions, and if you have a rear-engined car like a VW or Porsche, where the engine is swirling around in the dirt that accumulates at the back of the car, you need really good filtering. It's what put out the works 911s on the original London to Sydney, and caused a 356 to ruin the piston rings on the Peking to Paris in 2007, requiring an engine strip-down.

The basic pancake wire-wool filters are useless, while the old-fashioned oil-bath air cleaners, with a wire mesh content accepting the oil-breather pipe from the engine, are effective and easy to maintain in the middle of nowhere. A bigger item (the Capri V6 Essex engine has a nice flat filter across the top, which I've seen adapted to fit a Cortina) also makes sense.

A Porsche 356 heads out into the Sahara on the Around the World in 80 Days Rally. It has extra air-conditioning, having lost the rear window due to body-flex.

Chapter 3
Radiators and cooling

Cooling problems occur on just about all events. Even the Land's End Trial, run at Easter (so, that's surely set to run in blustery, freezing cold, rainy days) sees at least a few competitors suffer engine problems caused by overheating, as slow running uphill with mud being thrown up over the front of the car is a hazard faced by all. The Monte Carlo Challenge, which often found snow and ice on the higher passes and more than once encountered severe blizzards, saw radiators boiling over regularly. High mountain passes mean thin air, and engines always working hard, pulling in low gears. It's not the same scenery as the Sahara Desert, but similar problems occur – except that deserts mean hot ambient temperatures, so hot air going through a hot radiator is having to cool even hotter water, and must also succeed in transferring all that heat quickly and constantly. If not, the radiator blows its stack, and just where water is hard to find (Murphy's Law now having a hand in the disaster).

That's one aspect of the problem. The second, and absolutely the top priority if the car is to encounter rough roads or do any kind of off-roading, is the location of the radiator mounting itself.

On the Peking to Paris in '97, radiators split, broke their mountings, fell out of the frame and were held on by just the water hoses, radiator cores split up from their frames, old hoses let go in the heat, and jubilee clips lost their original grip, leaking water. We watched all of these things happen almost daily, and when the event ran again ten years later, the same problems occurred – with radiators shaking loose, hitting fans, or worse. Given that these problems had occurred previously, it really is a wonder that radiators gave rise to more reliability issues than just about anything else (well, exhausts coming loose and falling off may have been a bigger cause of aggravation, but radiators, on the Richter scale of driver 'quakes, are right up there). Sadly, history really does repeat itself, particularly on long-distance events.

We saw a 'wonder glue' from Robin Grant, who sadly died as this book was being compiled. Robin worked for the Renault Grand Prix team and fixed no end of radiator related problems with a dab of glue. But the fact remains, the problem was caused in the first place by workshops failing to appreciate that a vintage Bentley body pounding across Mongolia will flex and move – as it's designed to – in a way it will never do in the lanes of England. A tiny one-inch 'stay' or bracing bar was enough to do-in a Bentley that had consumed hundreds of hours of preparation, simply because the weight of the water (one gallon equals 10lb, and there is a lot more than a gallon in a vintage radiator) was pounding up and down amid a honeycombe maze of brass, until something had to give. The radiator breaks free, eventually.

SPEEDPRO SERIES

A simple rubber cotton reel Mini exhaust mounting. Also brilliant for mounting radiators.

The answer here is to re-engineer the mountings. The principle goes for an Allegro going on the Mongol Rally, a Triumph TR3 driving to Marrakesh, or a Peugeot driving to Cameroon – if it's going to be rough – potholes and broken tar being the worst kind of rough – and if you are facing dirt corrugations, the ripples of hard rolling-pins that cross the tracks in deserts, then the radiator will need serious attention in the mounting department.

Add rubber mountings for the radiator – those used for the Mini Cooper exhaust are ideal.

These rubber buttons have a thread on each side of a sandwich of rubber, which is 'waisted', curving in at the centre not unlike a Coke bottle. The shape helps the rubber flex slightly in the centre. Only an inch or two tall, these cotton reels are ideal in countering the damage done by vibration. They only cost a couple of pounds and this simple act of foresight was enough to ensure a Hunter's radiator could take a severe pounding, allowing the car to stride over the rough in the same manner that enabled Andrew Cowan to win outright the '68 London to Sydney (a victory only brought about by proving to be the most reliable car in the event; his Hunter was never the fastest on any particular day, being one of the least powerful cars, but it certainly proved one of the best prepared, and so the most reliable).

Had the vintage cars on a recent classic long-distance event adopted the use of the Mini Cooper exhaust rubbers, as suggested in the pre-event advice notes, a lot of these time consuming, energy sapping, demoralising problems could have been avoided. Unfortunately, it seems that many a driver just never made sure the advice notes reached the workshop, or if they did, that the advice was acted upon. Too many workshops are set in their ways, dismissing suggestions with "ah, we know about that, but we do it this way." If only those who do the screwing together of the nuts and bolts could also go and see how their handiwork suffers in the extremes of battle, then perhaps there might be less conservatism on the shop floor. Once in Death Valley, with water on the ground, it's too late to rue what might have been – but the frustration remains, because it's a problem that comes up time and time again.

Enough about mountings. What about the radiator itself? "What we need here is a bigger one ... a racing aluminium jobbie, right?" I wish I had a quid every time I heard that one. Wrong. There is probably nothing wrong with the standard radiator (unless you are in a Mini, and I will come to those in a moment). A bigger radiator is more weight, more water shaking about at the front end, more for the shock absorbers and springs to cope with. Apart from the water, the radiator will be a significantly heavier item. Is it strictly necessary?

The manufacturer designed the car anticipating that it would see hot climes, or slow driving down a farm track, or stop and start delivery work – the least of the worries would have been medium cruising speed with minimal gear changing.

So, now down to some pet likes and pet hates. It has to be stressed that there are opinions out there almost as strongly held as those regarding tyres when it comes to the finer points of cooling, and for every 'pro-this' there are those in an opposite camp of 'anti-that.'

Alloy radiators or old-fashioned brass? The restoration shop will probably talk you into an alloy version, as it is out to take your money at the end of the day – that's how they exist. However, is it better? I've rallied with both, and I've only ever had problems with alloy ones. I had an ex-works Triumph TR8, built by a team that could adopt a money-no-object approach to problem solving, and the full-house V8 engine had power to spare, but it ran as the hottest thing I've ever driven – even hotter than the ex-works Healey that blistered the fingers in my left hand whilst I was wearing a golfing glove, and melted the rubber of navigator Mark I'Anson's plimsolls, welding his feet to the floor. Did the Triumph, then, need its alloy radiator, lighter and slightly more efficient than brass? Maybe, maybe not – but I was the one who picked bits of plastic out of the gills when the electric fan disintegrated (it ran cooler the moment we threw that away, as the airflow suddenly increased), and I couldn't help noticing that the gills were all squished and bent from the damage of trying to digest the plastic particles.

Is there anything that can be done to improve a standard radiator? Vintage cars have header-tanks to take expansion. If you have one of these you are onto a good thing. The only downside is that there is a lot of water up there, and the seams need to be carefully inspected. The same principle is fitted to just about all modern cars today – an expansion bottle. Many classics, however, don't have one, and it's a good idea to adapt the radiator if

RADIATORS AND COOLING

you are asking a reconditioner to fit a new core. At the same time, they can make it 'export spec' with an extra core, or maybe give it a slight increase in height or depth. Also, insist it has the outlet pipe brazed into the neck of the filler for an expansion bottle.

What this does is quite simple: hot water will not boil away – it has to expand into the expansion bottle. As this is sealed at the top with a watertight fit, the principles of vacuum take over, and the moment the water level drops in the radiator, water is sucked back into it from the expansion bottle.

Blowing out the radiator water after fast running, down a motorway for example, when the car suddenly stops (e.g. at a garage to refuel) is best countered by simply leaving the engine running ... as heat-soak takes over, at the moment the fan and water pump stop running, heat has to go somewhere; it's moments like this that the water can't cope with a sudden increase in temperature and expands into the overflow bottle, or worse. Keeping the engine on tick-over helps to counter this.

There are a surprising number of cars on long-distance events that fail to

The radiator for the Allegro was sent to Serck to rebuild as a three-row core, increasing water capacity.

adapt to this little bit of security in the cooling department. On one Historic-rally there was an MG ZA Magnette – I came across it just as it had dumped the entire contents of the radiator out through the overflow pipe into the desert sand. We were a hundred miles from the nearest water supply. The metal expansion bottle of the MGC, or the plastic one of the standard Allegro, would have prevented the crew from losing time, besides getting very hot and frustrated.

Electric fans: some swear by them. Others swear at them. I'm in the latter camp.

The fan only does any good at speeds up to 30mph (50kph). After that, the radiator is kept cool entirely by the speed of the car moving through the air. In traffic, towns, slow running, a mountain where there are lots of hairpins, the radiator certainly needs some reinforcement. Fans then are a must on certain bits. But the manufacturer has also thought of this ... and built in a fan to cope. Even an engine tuned up 20 per cent can usually run pretty trouble-free with the standard fan. Electric fans mounted on the outside of the radiator block airflow, and can cause overheating at speeds higher than 30mph.

Other problems are simply to do with the way the fan itself is put together – they can fall to bits after a pounding down a rough road, and plastic chunks can get wedged in the radiator.

Radiators with more cores can be a solution, but can also make things worse. Minis are an example. A four-core radiator was fitted to late Minis, but if you talk to Mini owners, they all now rave about aftermarket two-core radiators, which are better because the airflow is less restricted and they have increased capacity. And no amount of tweaking of the rest of the cooling system will work if capacity in the first place is marginal (Minis, having side-mounted radiators, are always going to be a worry, as the ram of air going in to the front of the car doesn't hit the radiator).

Do oil coolers work as an auxiliary radiator? Don't even think about it. Just because it looks similar doesn't mean to say it is – oil will flow steadily through an oil cooler, thanks to open galleries; pour water through it and it will just rush through, not stopping long enough to sample the gasp of cooling breeze you are offering. A waste of time and effort.

Anti-freeze. Yes, or no? Er, half and

Plenty of louvres in the Sunbeam bonnet, but still not enough to avoid overheating in Morocco.

SPEEDPRO SERIES

Lifting the lid. Getting at the thermostat is easy ...

Removing the thermostat, replacing it with ...

... s simple racing MGB blanking sleeve, increases water flow, reduces temperature.

half, here, please. Do not go more than 50 per cent – it will not make it better if you do. Is it a good idea? Yes, in theory. Anti-freeze raises the freezing point, but also raises the boiling point, and a 50-50 mix means that your radiator now boils at 8F higher than before you poured it in. That's a bonus. Also, the green stains behind the jubilee-clip joints tell you where there is a leak – you probably would never have spotted the slow dribble before you put in some anti-freeze.

One other gizmo. Some engines run well by replacing the thermostat with a blanking sleeve, which is nothing more than a bangle of metal, like a bracelet, with a few holes in it. They cost less than a fiver. A-series engines built by the works rally team used this, as it got more water round to the number four cylinder. An alternative would be to just fit a thermostat that opens earlier and at lower temperatures, as it keeps the efficiency of the heater, but if you are setting out on a hard, arduous event, and know it's going to be very hot, then chucking out the thermostat knowing it's one less thing to go wrong is good thinking.

Chapter 4
Electrics

If your car coughs and splutters to a halt it is likely to be a fuel feed problem. If the engine cuts out immediately, look to the ignition system – loose connections on the coil, plugs, distributor (hairline crack in the cap), condenser, or rotor arm. If just one accessory isn't working, again, it could be a loose connection, a bare wire shorting to earth on the bodywork, or just a blown fuse.

Often older cars have been 'got at' and the wiring is a complete mess, with bits of wire added randomly and dodgy connections. Either that or it's old and the terminals and connectors have corroded. In which case, it may be easier to rewire, replacing the lot with a ready-made loom. However, unless this is upgraded, this will merely replicate the original and won't necessarily incorporate sufficient fuses. One solution is to fit the loom out of a later model; the version from a MkII Cortina is better than the MkI – at least it'll incorporate a fuse box, albeit a basic one.

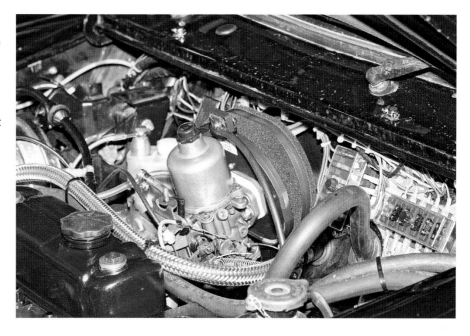

A beautifully wired Mini at the start of the Lombard. Note the fuses in see-through boxes.

The other rewiring solution is to use a rewire kit. The beauty of these is they're completely ready-wired.

Often, they're well designed – you have a central fuse board with a lovely bank of appropriate relays. In addition,

29

the exiting wires actually have their destination written on them. All you have to do is route it, trim it to length and stick a connector on the end – so simple it hurts. They're cheaper than you think too, with kits available from companies such as Pop Brown's and Langy's Speed Shop. A bit of research makes everything far easier and avoids fires.

When installing additional wiring, use the correct size to avoid the possibility of overloading and the wiring ending up a molten mess behind the dashboard. Take the wattage of the accessory to be connected – for example, a pair of spotlights with uprated bulbs would be 200 watts – divide by the voltage (12), giving a load of 17 amps. An electric fan takes a higher load, up to 30 amps. Where possible stick to the manufacturer's colour coding standard, as it makes fault tracing easier, and at least stick to the conventions of red for live and black for earth or neutral.

Use good quality switches (not the plastic bodied ones), and connectors (not the blue 'Scotch lock' ones which cut through the insulation to make a connection) – they do not do a 'proper' job and have loads of resistance. Invest in a decent set of wire strippers and crimping tools so that the wire cannot work loose or be pulled out by mistake. Ideally, all the connections should be soldered. Check that all the earth connections to the bodywork are clean and tight.

It is a good idea to include additional fuses, either in-line or through a fuse box, when adding accessories. To avoid a heavy load through a switch that might overload it (for example, extra spotlights wired through the dip switch), relays should be used. These act as a stand-alone switch wired into the main loom near to the item that needs operating. A small wire is all that is required for the low current that has to go through the dashboard switch, just enough to activate the relay – very simple, and economical on wiring, too.

How and when the accessory will be used will depend on whether it is wired into the circuits controlled by the ignition switch or not.

There are two main ways to switch the fans on – with a temperature sensor, usually mounted within the radiator housing, or manually, with a switch on the dash. Use thick wire and a relay, as it takes a lot of current. Don't wire it up so that when you switch off the ignition it is still running and you come back to a flat battery.

Check the routing of the wires to ensure that where it passes through or along any sharp edges of the bodywork it is either protected by rubber grommets or wrapped in extra insulation, or is kept in place by cable ties to avoid any extra movement or vibrations rubbing away the insulation and causing a short circuit

Under the dashboard, wire in a double socket, like a cigar lighter. These are invaluable for plugging accessories – the GPS, the light to wander outside with when changing a puncture in the dark, the small electric fan on the top of the dashboard that you hope will keep you cool when the sun gets really hot. There are small cheap self adhesive LED lights that are ideal for sticking under the bonnet and in the boot, which will help when scrabbling about in the dark, as will a small LED light that straps to your head.

Ensure the battery connections are clean and corrosion free – the latter ensured by smearing with vaseline – and that the battery is well secured in its tray. The sparks will fly if you hit a big bump and the battery becomes dislodged and touches the bodywork. Continual vibrations have been known to split a battery case open and cause everything to stop, as happened to Stirling Moss when crossing the Sahara in his Mercedes on the 1974 World Cup Rally. It was four days before a very dehydrated Moss was found and rescued.

Alternator, simple bracket and toothed belt.

ELECTRICS

Check the tension of the fan belt. Half an inch of play is right, slacker than this it will wear quicker and start to slip and not drive the water pump and alternator correctly. They do stretch and break, so take a right size spare (they are marked with their length in millimetres). Paddy Hopkirk lost a top placing on a Monte Carlo Rally when he discovered his mechanics had packed the wrong size spare belt.

A good dynamo will do the job, although it will only produce a charge above engine idling speed and its charging capacity is limited. Conversion kits to install alternators are readily available, which will give a steadier voltage of higher capacity at all engine speeds. Dynamos do have one advantage over alternators – they tolerate heat and vibration better. Note the elaborate trunking on the Austin 1800 in the photo, piping cold air to the back of the alternator – an example of the lengths to which the Abingdon mechanics were prepared to go to ensure maximum reliability on a non-stop, heads-down rally across remote regions.

We asked Simon Ayris of Rally Preparation Services for his tips on fault finding:

"Always start with the power lead of the troublesome unit, making sure there is a good power source to the unit with a test lamp. Use a good earth on the body of the car, with the lead on the test lamp. If that is okay, check the earth point of the faulty unit by attaching the earth lead of the test lamp onto the earth lead of the unit. If the earth is good, then the unit is faulty. If there is no earth, you have solved the problem!

"If there is no power to the unit, first check the fuse on its circuit. If power is only on one side of the fuse it will be the fuse itself, but something must have made it fail – either a bare wire on the fused side, or a fault within the unit itself. Further investigation will be needed at this point.

"If there is no power at the fuse, the switch that activates the circuit needs to be checked to see if there is power to the switch, and from it as well. If there is no power either side of the switch, check for damaged or broken wires using an ohms tester between components."

Here are Simon's six key points:
• A majority of electrical problems are down to bad earths, rather than bad feeds.
• Double-check all new terminals are safe.
• Insulation of exposed terminals – check this.
• Trip meters need to be wired separately – both feed wire and earth – to avoid spikes of electricity affecting the meter.
• Ensure all additional units have their own good earth, not shared with other components – especially trip meters.
• Any new units that draw large amounts of current MUST have relays fitted, and all joints should be made waterproof and dustproof with blobs of silicone gloop.

EMERGENCY ELECTRICAL KIT

• Pliers/wire strippers
• Small screwdriver
• Insulating tape
• Lengths of wire
• A simple circuit tester to check for current
• A selection of connectors (bullet or spade) and different rated fuses (glass, ceramic or spade) depending on the car
• Spare fan belt (pre-stretched)

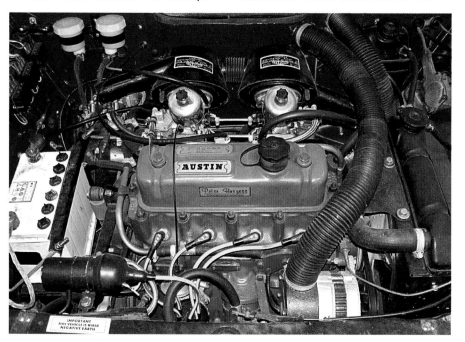

A works replica Austin 1800. Note the way cold air from the heater box is piped to the bearing at the back of the alternator, and the numbered plug leads.

Chapter 5
Suspension

Long-distance events see more problems concerning suspension than any other. The most common is not an absolute stopper – broken shock absorber mounts. If shock absorbers, or dampers, are uprated to go with stronger springs, then the area where they put stress into the bodywork needs reinforcing by a factor of four. Shock absorbers punched right up through the floor are a fairly common sight. Another common concern is that the bracket is forced off and the shock absorber is left trailing on the ground, wearing out the eye-bolt.

Springs that condense their shock into one small compact area, like the top of the McPherson strut, need considerable reinforcement. This was a problem when the Ford Cortina first came out in 1962; front wings were regularly punctured on club events. Either the springs were too soft so the suspension bottomed out, or the springs were too hard and punched up through the bonnet.

Ford Corsair about to be shipped to Cape Town for the three week Classic Safari – note the stout strut-brace across the top of the engine bay, and bath plug chain on the radiator cap.

Leaf-springs are simple to mend in the wild. After all, it's truck technology, and lashing up a broken spring is second nature in some parts. On events which use main roads that are well-trodden truck routes, such as the long hack across southern Siberia, garages are used to fixing suspensiona, and can make new leaf springs – sometimes old-fashioned means simple, so easier to repair.

The golden rule is to make the suspension firm and strong enough to put up with the constant pounding, day after day, over the hard corrugations that run across the track on desert dirt roads, and to give the car enhanced ground clearance to cope with the worst condition of all – badly potholed, broken tar. At the same time, it needs to be soft enough not to transmit lots of harshness and vibration to the rest of the car as it tries to shake itself apart.

One preparer I know says that as a general rule, suspension should be uprated by around 20 per cent – as a starting point – for rallying; here he was talking modern small hatchbacks. There are some cars where this would be rather excessive, and other large, soggy-sprung saloons where it's nothing like enough.

Interestingly, Ford avoided hard springs in order to win the London to Mexico. It wanted a car that could help

SUSPENSION

Riley leaf-springs appear well wrapped, but those retaining nuts pointing down look vulnerable, and there's no axle catch-strap to stop sudden snatch on full drop.

the crew combat fatigue, and a car that could take on all sorts of surfaces. The standard springs of 100lb rating went up to 115lb, but 2in longer (which gave a 1in increase in ride height) with substantially increased Bilstein shock absorbers. The softer springing was aimed at reducing the impact of shock and vibration – the car would shake apart if hard, Special Stage suspension had been fitted.

It is possible, with the right suspension, to drive over things such as speed humps at a decent lick (say 30mph) without a flinch (a Peugeot 504 could pass this test) – consider this before you think about better shock absorbers or estate car springs.

Check-straps under the axles, made of thin wire hawser with eye rings at each end as used in dinghy-sailing, would certainly help save the shock absorbers from the sudden snatch caused by a heavy axle when momentarily airborne and dropping down. Rattle over a stretch of potholes and the combination of a heavy wheel and tyre dropping down suddenly creates a snatch that places enormous strain on the shock absorber. You should limit the full extent of wheel travel with straps – it's a cheap and easy modification, yet it was ignored by most of the cars on the Peking to Paris. Interestingly, the Austin Allegro prepared for the Mongol Rally had rubber retaining straps on the rear suspension as standard. As they cost little and were light, spares were added to the list. Any form of retention to the front is going to be helpful, particularly if the system is a strut-type McPherson suspension.

Koni telescopic shock absorbers in their conventional oil form have a good reputation – you can mangle up the inside of a Koni and still they go on working. A good quality unit, they cost a bit more, but you get what you pay for. Koni has kept its drawings for old models, and if you can supply a shock absorber it can often rebuild it for you to the original specification.

Spax and Leda provide shock absorbers in uprated form for road events, and could be well worth checking out.

Springs and shock absorbers need replacing. Don't adopt the attitude 'it was all ok on the last event' – just because it looks all together, doesn't mean to say it isn't tired. Bushes in particular put up with a tremendous amount of wear and tear. A change to a more durable polybush, such as Powerflex or Superflex, might be prudent.

Bump-stop rubbers can play a big part in reducing the strain on springs and shocks – fitting bigger, stronger bump-stop rubbers, perhaps in polyurethane rather than rubber, can be very beneficial. The Morris Minor I took on the first Himalayan Rally had Land Rover bump-stop rubbers. When we lost a rear Spax shock absorber because the mounting broke, we were able to continue, riding on the bump-stops a lot more often, but still mobile.

Driving techniques can play a part. If you have some roll in the suspension, you can use this to good effect on corners. If you suddenly see a rock in the road you can swing the car hard over, so the side of the car that strikes the rock is being lifted up a little – a technique that is second-nature to those who regularly drive roads in Kenya. Hitting the brake pedal just before a washout or gulley only compresses the front suspension badly – you now have limited the amount of suspension travel. When you hit a hole or sudden boulder like this, you are actually increasing the amount of damage and shock going up through the suspension.

Adopt a technique of looking ahead and scrubbing speed off before the hole or boulder. If you're still going too fast, rattle over it – let the suspension do its

SPEEDPRO SERIES

Simon Ayris kept to period spec solid discs when rebuilding this Ford Corsair for the Classic Safari. This means it's eligible for a wider range of events.

It's obvious to all that the rear suspension on this Mongol Rally car has not been beefed up – it's carrying far too much weight and is on the bump-stops at the start. Trouble is only down the road – either chuck away some gear or uprate the springs.

best, and even if you have cracked your head on the roof and the navigator is swearing at you, the suspension has probably taken it all in its stride. If you can't find stronger McPherson struts, find some tubing of a similar diameter, perhaps a length of old piping, cut it down the middle and weld this to the side of the strut. Triangle shaped gussets at the bottom will stop it folding.

Finally, weight is the biggest enemy of a rally car. Dispensing with all non-essential luggage and cutting down on heavy spares is vital. You need a ruthless approach to this – first-timers always overlook it. Crews lugging suitcases across hotel car parks, with several changes of dress for dinner is all very well, but the poor suspension has to cope with it all.

You are just not competitive if you are carrying too much weight – sooner or later, the suspension is going to give out.

Don't forget to test it all well before the start, loaded with the kind of weight you are going to take with you. Even if you only run over sleeping policemen at the end of the street at 20mph, it will tell you something.

Suspension problems are the most common setback on long distance events, and while it's true the local workshops whose workers barely understand a word of English have seen it all before and know how to go about repairing things, prevention is by far the best cure. Uprating the springs, getting all the unnecessary weight right down, and testing it thoroughly beforehand is the best plan – and that means allowing enough time to ensure the car prep is not a rushed, last minute affair.

A lot can be learned on a dash down a local farm track ... tell yourself that Mongolia is eight long days of this, but worse.

Chapter 6
Brakes

Whether you are setting out on an all night clubbie or driving to one of the four corners of the earth, you will need good brakes and the peace of mind of knowing that all the seals and the master cylinder are within their prime, the discs are not all grooved and worn, and you have a good set of new pads.

A repair kit for the master cylinder is not good workshop practice – fine for the clutch hydraulics, but it's better to replace the master cylinder totally. Brake lines that are corroded, or rubber that is past its prime ought to be changed, and brake lines can be given some stone protection by running inside a long coil spring. The next step is a switch to steel braided lines, which flex a lot less than rubber under pressure, so give you a harder brake pedal.

Give the whole system a fresh supply of Dot 4 fluid. A word about brake pads, as more bull is bandied about on this subject than just about any other.

The temptation is to fit competition

Vented discs, callipers and pads – check the event regulations to make sure there are no restrictions. All-Historic events outlaw modern technology.

pads, since you'll be driving hard. But these work on heat ranges, and generally that's higher than road driving, meaning they don't work too well from cold. A

SPEEDPRO SERIES

Night events are punishing on brakes. Heavy braking for this Peugeot 205 on the Bro Teifi Rally in Wales.

race pad won't stop you much at all for the first mile or so, and you need to drive along carefully with your left foot dragging on the pedal to put some heat into the pads – best you avoid full race pads. Always use colder pads than you think necessary; you can go up a range if they're acting too soon, or wearing too quickly.

For general fast road use, start with slightly improved pads such as EBC Green Stuff, or the old favourites, Mintex 1144s. These are good for fast road use, and come into their own quickly with just a touch of heat.

Do not keep your foot on the brake pedal whilst waiting at time controls, as heat soak is transferred from the callipers, into the discs, into the wheel-bearing, which is taken up by the wheel-bearing grease, and so on ... heat has to go somewhere. Keep your foot off the pedal when the car is stationary, when no air is being forced around the hottest components.

Grooves and holes wear pads a lot more. In some cases, wear can be alarming. They might clear muck and water, but that's debatable – the advantage they are supposed to offer is probably overrated; the wear they cause on the pad face is beyond dispute.

Lastly, a word on vented, drilled and grooved discs. If you are running in an all-Historic rally, then you are restricted to solid discs, as only these were available in period. You may be allowed to upgrade from all drums to solid front discs, and there are kits available for older cars. Marina or even Escort MkIII parts bolt straight onto a Morris Minor, for example. Avoid the temptation to retro-fit modern technology. Check the regulations with the event organiser or the HRCR.

Chapter 7
Wheels and tyres

A fairly cheap item to obtain – and one of the biggest things that can transform your car ...

Is a nice set of alloys mandatory? Not at all. Standard steel wheels may be old fashioned, but they come with one bonus – small kinks and dents in the outer edge of the rim can easily be tapped out with a screw-driver. An alloy wheel is brittle, and hitting a pothole or brick in the road might shatter an alloy wheel, whereas a steel wheel would have survived at the same speed. Alloys are not necessarily as strong, and only a fraction lighter than steel.

Steel wheels can vary in strength. The toughest, most durable, and hardest to mangle are the old-fashioned Rostyle wheels made by Rubery Owen, fitted to older cars. They appeared on all sorts of things, from the Cortina to Sprites and Midgets towards the end of their life, and most notably on the 1600E, Rovers, Humber Sceptres and much more. Black with silver spokes, if

A works Porsche 911 ready for the start of the '68 London-Sydney – four spare tyres, an exhaust and air-intake piped up to roof-height for river crossings.

you spot these in a scrapyard you have struck gold.

Van wheels are often stronger as the manufacturer knows a delivery van is going to be kerbed and suffer more punishment. The van version of the Morris Minor, for example, is one of the toughest steel wheels made (and is handily one inch wider, so takes a 165-14 tyre nicely).

Wheels are 'unsprung' – that is, the suspension (spring and damper) is coping with the shocks of driving down the road, but carrying the weight outside of the spring – the whole wheel and tyre is not sprung. That's why racing cars want the lightest wheel and tyre – the strain of cornering is all the greater with heavier wheels. A wider, heavier tyre puts a lot more strain on the wheelbearings, and apart from the danger that the sidewall could catch the inside of the wheelarch in some extreme condition, like full bump while cornering (which can puncture the tyre – we see this often on long-distance events), the heavier tyre can contribute to axle failure.

The Suzuki shown here on the Mongol Rally has a halfshaft pulled out of the axle – one wonders if the wider tyre, being considerably heavier, contributed to this. So, is a wider tyre going to give you more performance? Not if you have to spend a day in a workshop hoping someone can patch it all back together.

SPEEDPRO SERIES

Jump for joy: an Austin A35 is hurled up a hillclimb on the Lands End Trial, fitted with the much stronger Rostyle wheels from a later Sprite.

Fiat wheels are not the strongest, but alloys would have shattered and tipped the car into a ditch.

Suzuki with broken halfshaft – an over-heavy, wider tyre added to the strain.

Choice of tyres is very much a personal thing, but here are a few pointers. Remoulds are popular in the British club scene – a firm called King Pin is long established, and remoulds these days are tested to the same standards as new tyres. These tyres are marketed by a company called Sportway for club rallying, and a Michelin MXT-patterned tyre using a winter Avon rubber with reinforced carcass promises to be a good budget tyre for road-based events, such as the Endurance scene.

Long-distance events should be free of any low profile tyre. You need every available millimetre of ground clearance, so, go for an 80 profile. You also need height in the sidewall because a chief concern is protecting the rim, giving you the maximum amount of air between the rim and the road surface. A good budget choice is either Avon's Supervan or Avanza.

An increase in pressure of 10lb or so could be a good idea if you are going over rough desert ground. For one thing you are keeping the blocks open, and so exploiting the tread pattern to its maximum whilst stiffening the sidewalls, therefore helping to reduce the chance of punctures from flying stones. Also, the tyres will run cooler if you add extra air, as well as cope better with unexpected potholes.

Talking of deserts, do pack an

WHEELS AND TYRES

The crew of this World Cup Rally Mazda have practised a drill to make wheel changes swift – they know where the wheelbrace and jack are, and where it all slots back after.

This Peugeot crew now regrets fitting wider tyres ... the last thing you need when it snows.

watched a tyre being stuffed with straw. The tube, getting fitted to a tyre with a hole in the sidewall, was a 14in being adapted to fit a 15in wheel ... there is always someone around who can come up with an answer to a problem.

Inner tubes can create punctures all of their own, entirely down to rubbing with the side of the radial tyre. Look inside any radial and you will see ribbing lines running across the inside of the tyre. On some tyres these are quite pronounced. You need to rub these down with a strip of sandpaper, as it's these ribs that cause friction burns when rubbing constantly into the inner tube. If you are watching someone fitting an inner tube for you, insist that they put in lots and lots of talcum powder. It can make all the difference.

Van tyres require weight to make the most of them (drive an unloaded Transit with eight ply tyres and you will soon discover that the tyres are not working). So, a light hatchback on a van tyre for strength is all very well, but there is going to be a downside – less wet grip and less grip on tight roundabouts. The rubber compounds are improving all the time, as vans are now speeding round the M25 overtaking everything in sight, so the tyre companies have to keep up with how the product is being treated.

There is a compromise – a tyre marked 'Reinforced' or 'Extra load.' These tyres are normally five ply, while the van tyre is six ply. Sometimes you have to look at the small print on the sidewall to see the difference – for example, the Avon Avanza can come in either reinforced or full van strength, and the tyre looks identical. The reinforced tyre has a touch more compliance, a touch more grip, and therefore better braking. If running a van tyre, you *must* inflate with a lot more air – 10 to 20lb on top of what you think is normal.

inner tube or two, and a puncture repair kit with really big patches – even if you can't do it yourself, there is bound to be someone around who can help. I was once walking down a street in the hill-station of Misoorie in Northern India and

Chapter 8
Bodyshells

If you are fitting uprated suspension, the next step will be reinforcing the mounting points, and strengthening the bodyshell where the stress loads have now increased. All bodyshells flex, to a degree. Vintage and Morgan bodies flex a lot. If you overdo the strengthening, say putting in long strips of seam welding, all that will happen is that the bodyshell will split open.

Tracking down someone who has prepared your type of car previously, whether it's a Cortina or a Peugeot 205, could help you glean a lot. While everyone has different ideas about engines, gearboxes, final drive ratios, brakes and suspension, the one thing you will agree on is whether the shell cracks up and where. For example, on Hillman Avengers and Sunbeams, the trick is to put a tube up through the inside of the windscreen pillar and strengthen the shell around the bottom corners of the windscreen – we only know this because others in the past have encountered problems.

Windscreens can pop out because the shell is flexing – a common problem. This is best countered by placing thin metal tabs on the top of the roof, coming over the windscreen rubber with perhaps a thin layer of cork or rubber touching the glass – it's just sufficient to stop the whole windscreen falling out.

The drawings of the Ford Escort bodyshell preparation later in this chapter show how extensive and ambitious you can be if you are deadly serious. These are the early drawings produced by Ford's Advanced Vehicles Operation in the early 1970s to help club drivers follow what the works rally team had learned after extensive testing for the London to Mexico. There are ideas and general principles here that could be applied to a number of other different makes and models.

The general principle that must be remembered is that if you strengthen something, then the stress has to feed in and be absorbed somewhere else.

Long-range petrol tanks are labour intensive to plumb in, and to manufacture in the first place. One way out here is to add or to adopt a tank from another model – the advantage being that you are using something that has been type-approved and gone through a manufacturer's crash testing. A common mod is to use an early Ford Transit tank (14 gallons). Other useful tanks well worthy of a look are the flat square tank of the MGB (10 gallons), or its bigger sister the MGC (12 gallons). You might have to get the maker to produce one in old-fashioned steel, as most are stainless steel these days (somewhat harder to weld if you are fitting retaining straps and brackets).

Another way is to take a strapped down metal petrol can. Paddy Hopkirk accessories, maker of roof racks and petrol cans, does one with metal expanding netting inside, which it calls 'Explosafe', and makes two gallons almost crash-proof. Demon Tweeks

BODYSHELLS

This works-prepared Austin Maxi entered the London to Mexico World Cup Rally in 1970 – without doubt the greatest road rally ever organised. It saw the longest timed competitive section of any rally ever; with a target time of 42 hours, everyone was hours late on the 560-mile blast to Bolivia. The Maxi of Rosemary Smith, with just 1500cc and 100bhp, was virtually standard – the biggest change was to weld the top half of the hatch to strengthen the body, leaving just a letterbox boot lid at the bottom, seen here revealing a massive long-range tank. How did she do? Rosemary finished 10th, and 1st lady, after an epic drive.

Simon Ayris fits Corbeau GTB seats and Luke belts to the Ford Ka entered on the Lombard by the rally editor of *Motorsport News* – not a simple car to prepare.

A bolt-in half-cage. This mounting is just as strong as welded-in mounting, but makes repairs easier.

offers for sale the full-size jerrycan at a good price, and also a half-size version, along with a separate filler tube to help you get it to pour out without glugging and splurging gushes of fuel down the side of the car. Forget plastic cans – the lids are never totally secure, as the plastic 'sweats' when hot and you can always smell fuel vapours around the lid.

Seating is one of the most important aspects of improving your bodyshell. There are specialists like Corbeau that offer a range of different seats, and if you are driving a long-distance event the first thing to remove from the list is a fixed-type racing hard bucket shell, usually made of Kevlar with minimal padding. You will yearn to move the seat forward or backwards or recline the backrest at some point on the trip, and you'll wish you had a more versatile seat.

However, if low-cost is the chief objective, then there is only one answer – a trip to the local scrapyard. Take with you a tape measure and consider what you have to do to remove the seat and how you are going to mount it (Corbeau can supply universal runners – just how universal they are is something that can only be discovered with experimentation).

Seats often rock from side to side which is more than a bit annoying – it's unsafe.

Stronger seat mountings are a must if you are going down rough roads – the seat brackets are taking the pounding of the car with the leverage of your weight, and unless it's good, something is going to give. Sideways movement of the seat in the runners can be annoying. The answer here is to fit a Jubilee clip right round the bottom of the seat and the runner, then screw it tight.

Rollcages come in all sorts of different shapes and sizes. If you are going to be driving quickly and competitively, then either a roll-over hoop with two backstays down towards the rear wheelarches or a full cage is sensible. This also strengthens the shell enormously.

Don't rely on the standard bonnet catch, which could free itself and fly up, smashing the windscreen as the body flexes. Add a leather or rubber strap or metal bonnet pins to be safe.

SPEEDPRO SERIES

Bonnet pins on the Classic Safari Corsair.

A tiny lever from a minivan poking through a radiator grille is how works Healeys did away with any possible problem of bonnet catches vibrating into the jammed position. This is simple, neat, and far more reliable than cable pulls.

Sheets of Thermawrap foil glued down under the Allegro carpet – it all helps to reduce in-car temperature.

Bracing the engine bay with a strut brace across the top of the McPherson struts is a good idea, and a range of braces can be obtained online or out of of the Demon Tweeks catalogue. Some cars (the first Cortinas) would collapse inwards without this.

If your car is on a 'one trip rally' because at the end it's going to be donated to a local charity, or you are on a very tight budget, you might be wondering if you can really cope with the amount of work that might suddenly be forced on you the moment you remove the sills. They could be hiding a total house of horrors, with corrosion that could call for days of chopping away and refabrication. If you think your car is well past its shelf life, and is going to do just a few club events and then be thrown away, or is not going to be a major part of your life, the solution as shown by Simon Ayris in the Allegro chapter is the answer. He just opted for welding in a strip of new metal down the *inside* of the car, stitch welding at the top to the inner sill, with the bottom welded to the floor. This strengthened the car a lot – on top of that, two tins of builder's expanding foam spprayed into the sills gave further reinforcement.

Getting a flow of air through the bodyshell and into hot areas ought to be on the agenda. This can often be done by playing around with the standard vents – look at the C-pillars in the Allegro chapter and see if you can remove interior trim or open them out. Roof vents like those on modern Subarus look totally naff on Ford Escorts and old cars.

Soundproofing is something normally ripped out of a rally car. If you are going on a long-distance event, however, you might like to add some. If you can stop the drumming of floor panels, and contain the heat from an exhaust system that is now tucked up close to the floor, you will be combating one of the chief causes of fatigue, which is why the Allegro floor was lined with foil and additional carpeting, glued and hammered to the floor.

Talking of heat leads us to vents – getting the air out of the under-bonnet area is as important as getting it in,

Piping cold air to the front brakes of a Riley Pathfinder ... good idea for a car with lousy brakes. But that sumpguard looks rather flimsy ...

BODYSHELLS

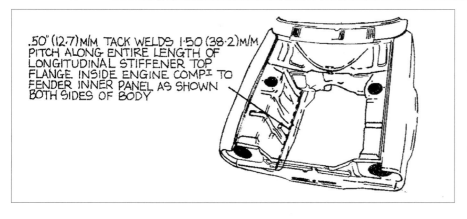

Works drawings of how Ford reinforced its London-Mexico Escorts. Much of this can apply to other models.

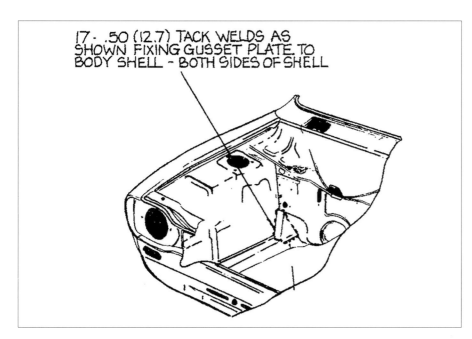

A simple triangle gusset greatly strengthens the weak area where the chassis rails join the bulkhead – also used by Maurice Gomm for Cowan's Hillman Hunter.

drawings produced by Ford and originally published by Jim Gavin. It's advice that not only pertains to Escorts – there are general principles here that can be applied to a very wide range of bodyshells with McPherson strut-type suspension.

The chief weakness is where the front of the car is designed to fold up in a head-on accident, where the chassis rails join the bulkhead. If it's a Vauxhall Chevette, Ford Cortina, Peugeot 205 or, as shown here, an Escort, this area will benefit from a gusset, a triangle of extra metal welded into this corner.

The welding of extra spots with gaps shows the kind of attention that was put into the works Escorts in the early 1970s. This advice was printed at the time to help clubmen go rallying, and was all based on the experience gained from having just won the awesome open-road event, the London to Mexico World Cup Rally. Ford shoehorned in a push-rod Ford Cortina engine, bored to over 1700cc, to win from a Triumph 2.5. The success gave us the car that took its name from the event, the Ford Escort Mexico.

Jim Gavin came onto the Ford tuning scene at a firm in Acton, called Supersport Engines, and drove the very first Cortina-engined Escort, a home-produced car which he drove on the 1968 London to Sydney Marathon.

He invented a process known as 'wedging' to strengthen McPherson struts. The struts tended to bend in Ford Cortinas in club rallying, and to solve this problem, the arrival of parking meters in Acton was manna from heaven – one dark night Jim went out and chopped a load of parking meters down, leaving the money box on the pavement and taking the tubing back to Supersport Engines, where he cut down the length of the tubing. This produced two half-moon strips of metal, which were wrapped

the simplest way being to raise the back edge of the bonnet by a couple of inches. Blanking off the sides of the radiator area can force more air through the radiator, which then still has to get out. The Healey rally cars had triangle slots cut in the sides of the wings which lowered temperatures under the bonnet dramatically; this was after discovering that vents in the top of the bonnet were not working – air was actually trying to get down in through the slots, not coming up and out (except when crawling through traffic or waiting at level crossings).

We have reprinted here some

round the length of the McPherson strut, and then welded in place. With a triangle of metal – a wedge – at the bottom, these proved more than strong enough for Cortinas to go club rallying.

The police were baffled as to why anyone chopping down parking meters would leave behind the money box – and Jim is still probably a wanted man in Acton. His mod would still work today; however, we think you should source some tubing without chopping down parking meters.

Spax, Koni and Bilstein all make heavy duty struts. There are two types of Escort shell, the MkI and MkII, the latter of which is the squarer, later shape. The MkI was given a 'Type 49' designation by the factory, with reinforced chassis rails running down the length of the car, which can be uprated yourself by extra welding. Additional spot welds, or brazing, makes the car stronger and stiffer, and therefore exploits stronger suspension better. Maurice Gomm of Woking prepared the early shells for Ford (and also the Hillman Hunter that won the '68 London to Sydney), and he used a mixture of conventional welding and also welding with brass, which is slightly softer and less prone to cracking, and requires less heat so is less prone to distorting the surrounding body area.

Here are a few tips from the original Jim Gavin prep-notes of 1971. We have used some of his drawings, as these could help guide someone preparing a similar bodyshell of the era, such as a Cortina or Anglia.

• Replace the rubber bungs blocking the floor with old pennies welded into position. The old bungs wouldn't last five minutes on a stage, and the car would be flooded the first time you cross a stream ...
• Put in a row of welds along both sides of the side rails or stiffeners under the floor. These welds should be 0.5 inches long, with a gap between each of 1.5in. A continuous weld is not necessary, and makes the car too rigid ...
• Gussets between top of chassis rails that bear the engine, and the bottom of the toe-board of the bulkhead, contribute a lot of strength – this also goes for Cortinas, Capris, and Corsairs (see drawings above).
• Remove the windscreen (because you are fitting a laminated one), and as the surround comes away it will be seen that it has been fitted over flanges that are spot welded together the whole way round the screen. Remove any wires such as the roof light's, and peel back the roof cloth lining. Examine the welds carefully, re-welding any that are suspect. Pay special attention to the bottom corners of the screen opening, where three pressings come together. These are from the front scuttle, the dash board, and the windscreen side pillars. Squeeze the three flanges tightly together using a vice grip, Mole Grip or similar, and then braze them together using neat welds, otherwise the screen will leak ...
• From inside the car, there is a vertical gap between each end of the steel dashboard pressing and the door side pillars. These should be cleaned with a wire brush, and then brazed their full length (about 4in each).
• There is no merit in having blobs of weld all along a seam, as brazing only works when it is inside or between flanges or seams. A proper welding job should be just about impossible to notice – not streams of snot!
• All Escorts have their rear shock absorbers inclined at an angle towards the centre of the car, and if serious competition or rough road work is contemplated they must be changed. Lots of motorsport accessory people do 'turret kits' these days so that shock absorbers can be mounted vertically. Fitting this is a job best done by an expert.
• A Ford Transit petrol tank holds 14 gallons and can be just about fitted into the boot of an Escort, using the original steel straps. A filler neck can be fitted through a suitable hole cut in the off-side rear wing, about 14in above the centre of the wheelarch.
• The petrol pipe on an Escort exits through the bottom of the boot floor – this is vulnerable, and while some people make a guard for this pipe, it really is better to make a pipe that will take the petrol from the fuel gauge unit at the front of the tank.
• All Escorts have a rubber pad fitted between each spring and the axle casing. This stops vibration and noise being transmitted to the bodywork, to a degree. They should be removed for competition work, as it's not possible to get the U-bolts really tight with these still in place ...
• Extra steel plates at the top of the McPherson strut as well as additional metalwork on the top of the inner wings is a must, along with a strut brace across the whole engine bay.

The MG ZR after the Bullnose Rally – note the way the air intake has been modified with the air cleaner now at the top of the engine, alongside the battery, away from the bottom of the engine bay where it will only suck in water. The strut brace across the engine bay is a vital bit of kit on these cars – springs have been known to hammer up through the top of the suspension turrets. The brace spreads the load, and stops the inner wings bending inward.

Chapter 9
Protection

Under-body protection should be a prime concern when preparing the bodywork. The first concern is to ensure there is a sumpguard to protect the bottom of the engine, and this can come in a variety of different forms.

One person found an old sheet of steel that was a discarded road sign from the A1 – it's now under the front end of a Vauxhall. A 6mm thick sheet of alloy chequer-plate is common practice, though not anywhere near as strong as dural (heat treated alloy, very difficult to bend), which perhaps is the ultimate.

If you are heading into the great unknown, then perhaps keeping to steel might be prudent, as it's going to be the easiest to repair, straighten out or weld.

There is no excuse for this if the wire-link joints are incorporated. Vibration breaks up standard exhaust mountings, with engine flexing adding to the grief.

Peugeot 205 on a Welsh road rally – it clearly needs that stout sumpguard.

SPEEDPRO SERIES

The mountings are a key ingredient. A sumpguard is only as good as the mountings on the chassis. Stout metal tubes, fixed onto something like Land Rover engine-mountings and bolted to the chassis rails would be good.

The space between the top of the sumpguard and the bottom of the engine pan fills up with thick mud, which dries with the heat of the engine, and worse, collects gravel and stones. This can then become a 'punch' that forces its way with ease through the bottom of the engine. Oil will now pour out in seconds, as it's under pressure, and the engine is totally ruined.

The way to prevent this from happening is to fill the gap with foam, a sheet of plywood, cork, or rubber. It's one of the cheapest things you can do – make up a block of cork by gluing together a few sheets of cork tiling from a hardware shop. Not only is it cheap, it's one of the most effective things you can do to protect your engine.

Everything that 'sticks down' underneath a car is going to be a drag when crossing sand, and is vulnerable to a bashing on rocky going. Spring hangers need triangular caps made to protect the front of the spring if you have old fashioned leaf springs at the back. Crossbeams need skidding. Perhaps a sheet of marine ply could cover up a vulnerable area of the chassis.

Petrol pipes and brake pipes ideally should be run through the inside of the car.

If time and budget constraints prevent this, then rather than ignoring the whole problem, come up with a way of covering the pipes up. Boxing them inside some U-bend shelving material or metal conduit is one way – not as good as running them inside the car, but better than ignoring the problem.

Look up at the car when it's on a ramp and imagine clouting items with a hammer – then consider that at 30mph. The force, and the potential damage, is far greater than anything you can do with a hammer.

This is better! Nuts under the spring hanger neatly capped, and bigger bump-stops on this Morris Minor, converted to Koni telescopics.

Four mud flaps are a good idea ... made of polypropylene, they don't come any tougher.

Added body protection from lining the wheelarches.

PROTECTION

Simple Dexion office shelving caps a petrol pipe – cheap and simple, and if the budget won't run to putting the pipes inside, why not?

A sheet of polypropylene covers a vulnerable petrol tank. Tanks at the extreme end of older cars take a pounding when dragged up steep inclines after river crossings.

'Roo-bar of a London-Sydney Austin 1800 to ward off kangaroos ... and Indian bicycles.

Rear differentials could often benefit from a skid plate – hard to weld to cast iron, but it can be done. You might be better off seeking out someone who has done this job before.

Polypropylene flaps for all four wheels are a good move, and why not run it high up inside the wheelarch to protect the bodyshell? Mud slides off polypropylene.

Petrol pipes might need protection, even under the bonnet, away from stones. Petrol boils at 40C, so wrapping petrol pipes in heat-reflective tape is a good idea.

Petrol tanks could often benefit from a degree of skid plate protection. You might think the manufacturer has tucked yours up out of the way, but the designer was not thinking of the steep inclines littered with stones when driving up out of a Mongolian river. A few layers of fibreglass bonded straight onto the tank is cheap, easy and light, as is a sheet of polypropylene. Even lino tiles have been known to find themselves glued here, and a sheet of marine ply has also proved useful. Creating a 'double skin' by using the bottom half of a tank from the same model of car is also a shrewd move. The one thing you can't do is weld a sheet of metal to the bottom of a petrol tank, unless you want to disturb the neighbours.

Protection for the front end – even the slightest bump from behind in an Istanbul traffic queue could wipe out all your lights. Fit an extra bumper, even if not a 'roo bar.

Sumpguards: the devil is in the detail ...

Here's a cautionary tale, and it's all true. On the Himalayan Rally with just about the smallest car, we thought we were doing really well, moving up from 5th to 4th, and then, after a long hard night trying to get past a Datsun from Kenya driven by Safari regular Jyant Shah, up to 3rd. You couldn't ask for better from the 1300cc Skoda Estelle.

Then we ran across a long wide river bed. Most of the water had dried up, so we picked our way carefully through large boulders, dropping into large holes now and then, inching our way slowly. Up and out the other side, and back into second and third gears ... but the vibration was enough for the sumpguard to come undone at the front.

A long thin sheet of Dural ran from the front bumper right under the length of the car, covering the engine and gearbox, emerging near the rear bumper. You can guess what happened next. In

SPEEDPRO SERIES

It can happen to anyone, including you. This VW Polo driver gets a helping hand on the World Cup to Morocco and back. He was soon righted and on his way.

Blinded by sun after a sudden corner, this MG ZR nearly dives off the edge of the Atlas Mountains in Morocco.

Could you cut this off? Can you undo it at the back? No, nobody can just saw Dural. And no, he can't get at the screws holding it onto the chassis, as the front of the sheeting has bent right round and is covering up the rear attachment. "Ah," says navigator Hywel Thomas, "I see it's held on with Allen keys! Have you an Allen key? If we can perhaps get in there, we can undo it all and catch up." Allen keys, countersunk, all nice and smooth ... back in the Czech workshop, that had made sense – beautifully engineered.

But on the edge of the Jim Corbett Tiger Reservation, the blacksmith was mystified, shaking his head from one shoulder to the next ... "Alan? Alan? Sahib, listen, I know Alan very well, I know truly ... he hasn't any keys."

The hard knocks lesson is that sumpguards need to be repairable. Choose a material that can be mended – if steel is what is going to be nearest to hand, forget saving a pound or two, make it out of steel. And sumpguards are only as good as the mountings. If one breaks, does it drag on the ground? Can it be got at? Why not add an extra mounting, each side, so if one breaks under the pounding you are giving it, there is now something else sharing the load. Finally, take a look at how your workshop has attached it. When it all goes wrong, they won't be around to nail it back on.

a jiff the front edge, having broken free, dropped down into the dirt, dug in, slid along a bit, then dug in some more, and then pole vaulted the whole car up in the air. Still attached at the back, the whole car was lifted right up – the rear now up in the air the length of the sumpguard, around 13 feet – and we crashed back down on the nose, with a long thin alloy sheet now bent double and pointing out the back. We dragged it along like that till we reached the next time control. Once on tarmac, the noise was incredible.

We found a village blacksmith.

Thanks to a strong car and a good sumpguard, it's soon on the road again.

Chapter 10
Stickers

Nothing lets down the finished look more than badly fitted stickers and door numbers. It is possible to get them to look as though they were air-brush painted, without a bubble or ripple. Here are a few tips – but it cannot be done single-handed. For the best results, you need a critic, someone who can stand back and is not slow in coming forward with the 'that's not straight' verdict.

NUMBERS

It is very tricky to put a number on straight. Use a small bar of soap to put on lines which you can butt the number up to, using horizontals first – the top of the number 5 or 7, for example, or the bottom of a 2. Get that down and the rest underneath drops in the right place. The trickiest of all is the number 8, and 9 can easily look lopsided.

DOORPLATES

Again, someone who can help you keep the two sticky bits from touching each other and ruining the whole sticker will help, so doing this single-handed is best avoided. Before removing the back paper, soak the whole door area with warmish soapy water – small suds created by lots of Fairy Liquid is ideal,

Robin Stretton, the 'sticker man,' shows how it should be done in his Citroën C15 van.

SPEEDPRO SERIES

Ensure there is plenty of water slopping around. Peel back the paper, ensuring your fingers are not removing the thin layer of glue by touching this tacky side. Place roughly in place, the water now ensuring you can slide the sticker or rally plate into place. Stand back to ensure you have the bottom edge horizontal. Go for the sill line as the guide, not trim lines.

Now use a plastic hotel doorkey, old credit card or old wiper blade, and rub from the centre outwards. The water and bubbles will disappear. This is a time consuming little chore and not something to be rushed. Keep pressing out from the centre, and finally change to a dry rag, as this picks up the moisture you are pushing out.

You might fail at this, and end up with bubbles. These can only be removed using a needle ... a small prick is all that is necessary to get things nice and flat. The moisture does dry – don't ask how it works, but dry it will.

Don't feel tempted to put on large door plates or stickers without soapy water. However good you are, you will get a crease or a bubble.

CREW NAMES

Even if the event doesn't make a requirement for these, putting your crew names on the front doors, or on top of the front wings Historic rally style, is a good move. Alison at Coba Graphics (see list of suppliers) will print these on strips and post to you as computer-cut artwork, so all you have to do is rub down onto the clean paintwork. Don't feel tempted to put blood groups on a car; no doctor is going to fill a half-dead patient with fresh blood on the basis of what is written on the side of a car. Advertising the gory side of our sport is something we should all avoid.

Putting on stickers and numbers on a windy day is not helpful - fine dust can cause havoc. Equally, cold days make things harder. If these conditions cannot be avoided, one tip is to place the stickers and numbers on a radiator before pulling back the paper backing – this way you are less likely to crease the plastic.

Removing stickers without damaging paintwork can easily be done if you first of all soak the area with a kettle of boiling water and allow a minute or two for the glue to soften up – it then peels back without a problem. Hairdryers may also be good for this.

Finally, don't go for sponsorship on a UK road event – stickers are forbidden by the MSA.

Prepared and entered on the 2002 World Cup by the Mercedes Dealer Team, a Merc A class with a fair number of stickers, seen here in Albania, driving London to Athens.

Chapter 11
Navigation equipment & gadgets

It's the navigator who takes charge the moment the car leaves the start. He or she decides just about everything – the pace, directions, when to stop, when to book into time controls, when to take on petrol or snatch a coffee, feeding the driver with the essential information as well as peppermints and Red Bull. All this decision making is best taken as the responsibility of one person – without interference from a driver, who ought to have sufficient confidence in the office administration without having to keep demanding "are we nearly there yet?" or the even more annoying "are you sure?"

When pressing on, the driver has a steering wheel to hold onto. Following a route on a map cannot be done bouncing around, and so to do their job properly, the navigator or co-driver needs to be belted in, with a lightweight full harness, a simple push-button buckle like the Luke, a footrest bolted to the floor, a grab handle on the door, and additional padding at shoulder and knee

A footbrace aids co-driver comfort.

levels. Seats that recline and tip forward make access so much easier. Roll cages comprising door bars and diagonal bracing across the back make living in the car that much more inconvenient, but the balancing act here is one of convenience against safety.

The dashboard in front of the navigator can look rather like something out of a black and white wartime film about a mission with bomber command. Depending on the type of event there

Mission Control, we have lift-off! A Mini Cooper dash – pity about the wires hanging down where the navigator grabs something while getting in and out, and knees close to the sharp edges of that dash could be vulnerable.

maybe one or more clocks, a mechanical (Halda) or digital (Terratrip or Brantz) tripmeter and a GPS (Garmin) display. The latter can prove vital in remote regions where there are no roads shown on the map and you have to rely upon the compass needle displayed on the GPS screen to guide you from one

SPEEDPRO SERIES

Halda Twinmaster, next to the Lancaster Bomber clock, with an older Halda Speedpilot underneath, complete with leather key fob. Riley ready for the Monte.

Morris Minor dash, with a Halda mounted in the centre so that it can also be read by the driver, a good stout footbrace, and pencil holders on the rollcage, left.

waypoint to the next. Pens, pencils, rubbers and spare light bulbs are also affixed to the dashboard – Velcro is ideal for these little jobs.

Two low-cost gadgets will prove most handy. Pre-prepared route instructions, place names and road numbers, written on postcards and placed in a frame that is mounted on the top of the dash and lit by an aircraft-style pea light. This enables the driver to follow the route whilst the navigator enjoys a sleep.

The other useful bits of kit are a flexible map light and possibly a Poti, an illuminated magnifying light, for finding the correct route when the maps are small scale or the lighting is poor. A long-distance rally car needs good interior lighting. You are almost living in the car, so check out the range of caravan strip lights.

Perhaps the cheapest items of equipment are the boards for securing timecards and maps. These can be made of corex or stout cardboard, ideally firm enough to bulldog clip the paperwork, but not too strong to do you an injury if the driver suddenly decides to explore the undergrowth.

Traditional Avanti map light. Always tape a spare bulb somewhere handy.

Whatever the event, whether reading from a map or a traditional Tulip-style road book, where interval distances are given between junctions as well as

NAVIGATION EQUIPMENT AND GADGETS

The Terratrip has lovely big buttons, making it easy to hit the right one when tired and on a bumpy road.

A novice navigator gets set for the Drystone Rally in a baby Fiat ... novices on British club road rallies have extra plotting time to get their paperwork in order.

total distances between checkpoints, then a separate tripmeter for the navigator is essential.

Traditional, 'original' tripmeters were either the clockwork Speedpilot, showing average speed, or the Twinmaster, showing distances. Both were made by the Swedish taxi-meter firm of Halda, having been invented by Dutch rally driver Maurice Gatsonides, the Gatso of today's speed cameras! These were driven off the speedometer cable, and calibrated using different size brass wheels. A bit fiddly, but great if you can find one. Second-hand they now command premium prices, so are hardly low-cost. They were replaced by the electronic digital display tripmeters, working off a sender unit attached to a wheel hub – much easier to calibrate and much cheaper to source and purchase: around a hundred pounds for a basic version. Both Brantz and Terratrip have their devotees. Basic models that do not compute average speeds may be permitted on some Historic events, but not all.

Now, the GPS receivers provide all the functions of distance, actual and average speeds (miles or kilometres) and exact location anywhere in the world, in a much more compact unit at very reasonable prices.

All the paperwork, copies of personal documents, roadbooks, maps and timecards should be stowed safely and securely so that they come readily to hand. Additional map pockets and extra cubby holes are always useful for that tub of wet wipes, a torch and camera. Everything neat and tidy ... a place for everything and everything in its place. When the driver demands a peppermint, he's only going to moan all the more if you then start grubbing around under the seat.

Sources for all this navigational can be found in the 'Useful contacts' section at the back of this book.

Chapter 12
Safety

If you take part in a UK road event, either a road rally or a navigational rally, *no* mandatory safety items are insisted upon; you could drive in an utterly standard hire car. However, just about everyone fits full-harness belts and carries a fire-extinguisher and first aid kit.

If you are going abroad on the Mongol Rally, Plymouth Dakar, or the Scally Rally around Europe, there are also no safety regulations – all safety matters are left as the personal responsibility of the entrant.

Belts can transform your comfort and car control. A navigator doesn't have a wheel to hold onto, and is always being thrown around, side to side and up and down, and 20 minutes on a really twisty bumpy series of lanes would be enough to cause even those with a stomach as tough as a boiler to feel car-sick. A driver is *considerably* quicker in a full harness belt, the sort that looks as if it's from an aircraft pilot's cockpit. All the major belt manufacturers – Luke, Williams, TRS, and the French Sabelt concern – make the lighter road-going model at less than £50 a set, and these are what the author recommends for competitive road events. We have used Luke belts, good value with beautiful stitching quality.

Also, get a fire extinguisher – one large enough to put out a fire (AFFF, 2-litre) – and mount it so that both crew members can easily reach it.

A rollover cage, either just a hoop or a full cage (which can have front legs coming down the inside of the windscreen pillars, side bars over the tops of the doors, and even bars across the door openings) is not mandatory on road events – it's a matter of personal choice. It is fairly easy to bang your head on all this tubing, even in a minor incident, so some dense foam covering is a must – Trident Supplies of Silverstone sells this, as does Demon Tweaks.

You need to see that items in the car are securely strapped down – this is easy to overlook, but essential. Spare wheels are best in the boot or hatch area, along with jacks, wheelbrace and tools. On the first Himalayan Rally, a United Nations aid worker died driving a Land Rover when he hit a gulley in the road and was struck on the back of the head by a jack handle. It was his first big event.

Keeping boot weight as light as possible. Here, the tools and spares are in neat trays that cannot fall apart, and the jack is well strapped down. Note the big fuel tank placed forward over the axle.

Chapter 13
Keep it going

THE TEN MOST COMMON FAILINGS
1) Shock absorber mountings snap.
2) Springs break.
3) McPherson strut comes through the bonnet or turrets at the top of the wings; inner wings buckle.
4) Radiator mountings break.
5) Engine mountings break.
6) Stones lodge between top of sumpguard and sump, holing sump.
7) Electric fans stop working, or fall off.
8) Water boils away, no expansion bottle to catch it.
9) Exhaust falls off.
10) Battery container breaks.

THE MOST VALUED ITEMS IN THE TOOL BAG
You should never leave home without gaffer tape, nylon tie-wraps, tie-down or bungy straps, Bar's Leaks (for radiator), plastic padding or other epoxy resin, a can of WD40, coat hanger wire, garden fence-style link chain, Opal Fruits (Starbursts), a pair of pliers, a Mole grip-style adjustable spanner, and a decent hammer.

THE BEST ROADSIDE BODGES TO HELP YOU SURVIVE
1) Opal Fruits (Starbursts) to plug a split in the petrol tank (then cover in Araldite).
2) Old coke or beer can, cut open with scissors, wrapped around a hole in the exhaust pipe, clamped with a Jubilee clip.
3) Chewing gum then Araldite for a hole in the radiator.
4) Wire from a fence can replace a broken throttle cable.
5) String through the side window to work a broken windscreen wiper – until the navigator gets bored.
6) Fix a broken fan belt with a leg from a pair of tights – a fishnet stocking is even better!
7) Always have a spare ignition key taped to the inside of a bumper.
8) Superglue a crack in the windscreen the moment it happens – this can stop it spreading.
9) Fuel boils at 40C – if a misfire is being caused by heat soak into the carburettor, glue layers of BacoFoil under the carb and lag the fuel lines with foil.
10) If an SU fuel pump stops ticking, belt it with a shoe.

Chapter 14
Twenty bargain project cars

FIAT PANDA & UNO

In the bargain basement category, nothing is much cheaper than an old Fiat, certainly when it comes to refurbishing a car with new discs, clutch and the like – all cheaper than chips. The Fiat Panda and the Uno are the staple diet on events with a 1-litre rule, like the Mongol Rally.

Replacing tired soggy springs, clutch and brakes can all be done on eBay. The original Panda came with rear leaf springs, which is a great bonus, being so much simpler to repair in the wild, but finding a good example is not easy. The car was intended as a rival to the VW Beetle/Renault 4/Citroën 2CV, and everything is basic.

The Panda was introduced in 1980 and used a lot of Fiat 127 parts, including the push-rod engine. It was phased out in 1996, after Seat made a similar version in Spain, when the car was replaced by an all-new and more sophisticated model. Pay particular

This Fiat, with no dents or scratches, 56,000 miles, a full service history, and a sound floor and sills, rests on four perished tyres in a front garden, spotted in an Oxfordshire village ... the kind of car you can buy for £150 on eBay. New discs, clutch, and radiator will cost a few hundred pounds, and then take you to Mars.

attention to the tin worm, as all old Fiats have a poor reputation for rust prevention. Dashboard squeaks are common on all – glue all the plastic together with blobs of Araldite.

The later Uno with the all-alloy engine, called the 'FIRE' engine, is a great revver and makes the car fun to drive. The whole car is very maintenance friendly, with everything easy to get at and work on. The box-like shape is surprisingly effective, and had an aerodynamic Cd figure of 03.40 that set new standards of efficiency when introduced – it probably means that if you put a rack and spare wheels on the roof, it will still deliver 35mpg.

Talking of wheels, the standard steel wheels are not very strong – take two spares on 80 profile tyres, or ask Weller Wheels to make you a set.

NISSAN MICRA

At the time of writing, an astonishing 30 per cent of the total number of original box-shaped K10 Nissan Micras are still on British roads. The later dumpy, rounded Micras are even more plentiful. A favourite of driving schools, old ones may have 100,000 miles or more on the clock and still run smoothly, capable

TWENTY BARGAIN PROJECT CARS

Nissan Micras may be boring to drive, but they have super-quality electrics and top build quality – easy to transform into a rally car.

Adrian Booth takes an unusual line in his 205 as he storms up Bishopswood on the Lands End Trial, scoring a penalty-free run. Note the lack of full harness or roll-cage.

of pootling along and holding up the queues at every roundabout.

The Micra has a brilliant reliability record. The electrics, all the little fittings as well as the wiring loom itself, will never fall around your ankles or spark in the wrong place – or stop working, which you can't say about French rivals like the Peugeot 205 or 106.

The rear suspension will take a Group Four Escort rear shock absorber as a direct replacement fit, but why would you want to? It's going to be far too hard.

As the car was used in British club rallying, finding a sumpguard and a few rally bits ought to be perfectly straightforward. Journalists all describe driving a Micra as 'dull' or 'uninspiring,' but what cannot be denied is that they are brilliantly well made – and tough with it.

PEUGEOT 205

The 205 shook the world of rallying in the 1980s, at club level as well as at the World Championship, and if you can find a good rust-free one, you have a good car – easy to work on, with all the bits and pieces you need readily available. Finding the right shock absorbers or springs is easy, as someone else before you has done all the homework. The 1.6 GTi and its 1.9 bigger brother hold their price, and are now almost collector's items, but the smaller versions can take items off sister cars – springs from a diesel, for example, would uprate your suspension by 25 per cent in a jiff.

The 106 is a front-runner in the 1400cc-only Endurance rally scene in Britain, but those who compete against them in rival Vauxhall Corsas and Astras reckon the Vauxhall has stronger transmissions and can take more punishment.

If you are on a long-distance event going into former French colonies such as Morocco and Senegal, a Peugeot makes a lot of sense.

ROVER 25

The Rover 25 and older versions with the K-series all-alloy twin-cam are a popular and cheap choice in Endurance rallying, and Adrian Grinstead took the one seen here on the Tunisia edition of the World Cup Rally in 2003, finishing

Adrian Grinstead and Nick Mason finished second overall on the London-Dakar World Cup Rally, possibly the best ever international rally result for a Rover.

second overall – that must be the best ever international rally result by anyone in a Rover!

With the 1400cc engine there are two states of tune – 84 and 103bhp. The only difference is the throttle body, so swap to the bigger 52mm one and it will be the

57

SPEEDPRO SERIES

cheapest 19bhp ever. Avoid the earlier 200 with the XW chassis prefix built from 89 to 95. They are heavier, and the rear suspension is made from chocolate.

The Turner brothers who run the Rover Centre in St Albans have chalked up a string of outright wins in British Endurance rallies with their cars, and they now prepare Rovers for others.

They have also won events in the much-maligned 1400cc-engined Rover Metro. The Metro is considerably lighter, but runs Hydragas suspension – the fact that this is virtually impossible to repair by a wandering nomad ought to rule it out of any consideration for a long-distance banger rally.

PERODUA KALISA

This is the cheapest new car on the market. A brand new car, with manufacturer's guarantee, will give you change out of five grand.

If you have never heard of it but think it looks familiar, that's because it's a direct descendent of the smallest Daihatsu, and uses the same three-cylinder engine. It's dumpy and brick-looking, with an old-school Mini-style upright windscreen, but under what looks like a flimsy skin is a car with a dual personality. It's actually as tough as old boots, and great fun to drive.

Get the cheapest version, which comes without power steering – the tyres are so skinny it doesn't need it. Memories come flooding back of what it was like to hack a Mini down a country lane with lovely, pin-sharp steering in your hands, an agile handler that will respond to the smallest of wrist inputs, and with just enough power to convince you that your right foot can contribute to the way it hustles through corners.

Second-hand, they cost peanuts. Two things transform the car – underneath, you will notice ladder-like old fashioned box-chassis sections

The Perodua of district nurse Karen Young tackles a night section in Greece on the London-Athens World Cup Rally – it is a very similar car to the Daihatsu Cuore.

under the floor. You can in effect double-skin the floor by welding over these with panels, and filling the inside of the boxes with foam. You can add slightly longer springs, an inch at most, uprate by a mere 10 per cent, and uprate the shocks – you will have to go to a specialist like Leda to do this, and *insist* that it doesn't make them too hard. Or, fit MkI VW Golf inserts at the front, and use Perodua Kenari at the rear. You then have a brilliant little road rally car that is a lot of smiles per mile.

The three cylinder engine has a distinctive double-beat note reminiscent of a Porsche 356. Treat it to a forward facing, ram jet K&N or similar cylindrical air cleaner on a long hose, which will ram air in the faster you go, and give it the normal treatment of a sumpguard, waterproofing and flexible joints in the exhaust, and you have a great little car that is a lot of fun to drive. Tough enough to drive to Mongolia? Absolutely. If 1-litre is the maximum, this will beat more than most, providing you don't overload it with lots of luggage.

District Nurse Karen Young (no relation) uses one to make her daily rounds – at weekends the same car goes rallying. Karen has won the 1-litre class

on the four-days-and-nights Lombard Rally. Sophie Robinson took a Daihatsu (prepared by Paul Jackson of Didcot on behalf of the manufacturer) on the London to Athens World Cup Rally, and won the Ladies Prize and the 1-litre class.

David Johnson (email: bridgemotorsport@tiscali.co.uk) has turned these into cheap rally cars, and may undertake some preparation work on Perodua and Proton.

PROTON SATRIA

Like the Perodua, it comes from Malaysia and has absolutely zero badge appeal, but the Proton has won over a strong following in British club rally circles. What you get is a car with a lot of Mitsubishi heritage – the 'old style' Satria is proven to be one of the toughest, best-made small cars on the road.

What transforms it into a good club rally car, however, is that Mitsubishi Evo 3 suspension goes straight on, and totally transforms the chassis.

As a second-hand car the Proton has lost all its residual value, parts are cheap and plentiful, and bolt on bits from the likes of Harry Hockley are off the shelf.

Most of the British club events are 1800cc, but there have been good showings in the sub-1400 Endurance rally category, and in David

The Proton Satria has proven to be a strong and easily prepared road rally car.

TWENTY BARGAIN PROJECT CARS

Johnson's hands one finished third overall on a World Cup Rally. Apart from a determined driver, the car did well because it rides the bumps, has excellent handling, and has considerable reserves of strength.

HILLMAN HUNTER

The Hillman Hunter has won three big international rallies in its time – the first London to Sydney Marathon in 1968 with Andrew Cowan, the second was the longest rally ever, the Around the World in 80 Days in 2000, driven by Freddie and Jan Giles (which had just 80bhp at the wheels, but was faultlessly reliable), and the first Plymouth Dakar in 2003.

The car Freddie Giles drove had won the Ladies Award on the Peking to Paris in 1997, so it's probably the most well-rallied Hunter of all time. The car ran on two ex-Rover SD1 1.75in SU carbs, an original Holbay Hunter exhaust manifold, and a Transit van 14 gallon petrol tank in the boot. Suspension was estate-car springs at the back, and 140lb springs at the front, with uprated dampers. It was found in the classified section of *Classic Car Weekly* for £350.

The shell is enormously strong, as good as a Peugeot 404 but heavier than a MkII Cortina – most of the shell strengthening mods in the Jim Gavin notes for Escorts apply to the Hunter, including the most important of all, a gusset between the top of the chassis rails and the bulkhead. Rostyle wheels from any Ford will fit, and are the strongest steel wheel you can find. A well-built Hunter might not be the most powerful car, but you can thrash one all day long – nothing drops off.

AUSTIN 1800

Having mentioned the Hunter, meet its arch rival, the car that came second on the original London to Sydney. The 1800 is big, comfortable, front-wheel-drive with really good steering, a little low-geared but full of feedback – a well set up 1800 can be thrown around just like a big Mini. The shell is enormously strong – the works team cars had no extra welding on the shell at all.

The works cars used mostly bits from the parts bins of Leyland, and it's easy to transform one into a surprisingly good performer – you need to fit the rear suspension carriers of the Australian pick-up version, called the Ute, which the Landcrab Club can help with, along with Koni rear shocks and big bump-stops. A comfortable car helps beat fatigue, and few cars from the sixties can come close to the ride comfort of the old Landcrab, (so called because of its off-centre front headlight surrounds).

1800 Landcrab – long-distance cars don't come more comfortable.

You still see ex-works cars in Historic rallies today – one survivor of both the London to Sydney and London to Mexico Marathons won the Monte Carlo Challenge.

Cheap to find, the old Landcrab is still one of the most comfortable long-distance classics you can get. And it's different. Ken Green of the Owner's Club can help you.

SUZUKI 4X4

Getting hold of one of the early 1-litre little SJ 4x4 Suzukis is now rather difficult, because every year lots trek off

Replica of the Hillman Hunter that clinched the '68 London to Sydney.

A Suzuki at the start of the Mongol Rally. Don't you just love the spray-on mud on the tyres for that essential Hyde Park look?

SPEEDPRO SERIES

on the Mongol Rally to be given away to locals. Pros: good ground clearance, good grip, versatile, and 4x4 means you have no excuses. Cons: reputation for dodgy handling – they can fall over.

Don't fit wider wheels and tyres – it places more strain on the driveshafts. Get used to the limitations of driving a short, tall car, and don't overload it.

The Jimmy is an altogether more sophisticated car with coil springs – heavier, but it comes with a 1300cc engine.

New battery, waterproofing, and rally exhaust, all as suggested in this book, should be on the build sheet. Stronger engine mounts, decent shock absorbers, and better springs can all be obtained from Rhinoray. Check out www.rhinoray.co.uk, or call up the workshop in Woodingdean, East Susssex, on 01273 696796 – it also does a range of engine swaps.

VAUXHALL CHEVETTE

The Vauxhall Chevette was truly groundbreaking when it first came out. Here was a hatchback, with a rounded aerodynamic nose, nicely proportioned. It's rear-wheel-drive, has a strong axle (it can take 100bhp), is nicely located with a three link setup, and the differential is retained by a torque tube.

The bodyshell benefits from strengthening around the gearbox mountings, and the dome inside the engine bay which caps the top of the suspension can be double-skinned by fitting a repair item, so if you are uprating the springs and shock absorbers, this is a must.

The van and the saloon-booted versions have upright rear shock absorbers, which is great. It is possible to convert the hatchback to this, but you ought to track down a copy of Andrew Duerden's *Vauxhall Sportpack Manual*, as this has all the drawings and instructions.

Viva GT – a gutsy and underrated car, seen here in Morocco's Sahara.

A Cortina 28/36 carburettor goes straight on. So does a Stromberg 175CD, and an SU can also be made to fit – changing the carb will get several extra bhp. An extra eight bhp comes from a ported head with 36.5mm inlet valve. If the compression ratio goes up from 8.7:1 to 10.0:1, then the benefit is 10bhp. A few more horses come from a better manifold.

Rostyle wheels fitted to some later Chevettes have a different wheel centre and may not be interchangeable, but are about the strongest steel wheel you can get. Chevette wheels are 5j, and Cavalier GLS wheels are 5.5, and interchangeable.

Check out www.europerformance.co.uk, as it lists Koni shock absorbers and disc brake conversions for Chevettes.

The car is a delight to drive and fun, nicely weighted steering is full of feedback and the whole car is wonderfully balanced – step out of today's modern breed of overweight hatch with power steering and you suddenly discover a lost world. The Chevette was ahead of its time, now cheap to buy and simple to put right.

A rarity is the 2-litre GT Viva, but a mint condition car in low-mileage original spec will cost you no more than the price of overhauling a Salisbury axle in a rally Escort. The Astra GTE is often a front-runner in British road rallying, and the Corsa and Nova do well in Endurance rallying or any other class up to 1400cc.

MERCEDES

If you are driving on the Plymouth Dakar and dumping the car in a sale for charity at the other end, this is the make all the locals will want. Treat it to some shock absorbers, a sump shield, battery strap, and off you go – simple as that. Estate car springs will lift it up. Don't be too put off by automatic gearboxes, they are pretty bullet proof – new fluids and a filter wash-out ought to be top of the job list.

A tough old Merc taxi wins its class on the Peking Paris; the cheapest car on the event.

TWENTY BARGAIN PROJECT CARS

Peugeot put its lions into Africa with the 404 and 504, winning numerous events.

A small Volvo with a 2-litre engine storms up Nailsworth Ladder on the Cotswold Clouds Trial.

PEUGEOT 504

If you are going into what was French-run North Africa, this is the car to have. Most have rotted away, but they are enormously strong, simple, and very comfortable. The 1800cc engine is more sweet-revving than the 2-litre and just as quick. The rear suspension can come with an independent or solid beam axle, and there are two wheel centre sizes, as the estate car wheels are not interchangeable with the saloons. The estate springs can go on the saloon and really hike it up, but you get odd-looking positive camber at the front, a bit like a Vintage Bentley.

I've owned over a dozen of them, and rallied a pick-up version (homologated into Group B as it's a two-seater!) on Yorkshire's National Breakdown Rally. If you are looking for a cheap car for rallying in Africa you can do much worse than choose the model which built its reputation on African Safari rallying.

The best tyre to go for is the winter van 185-14, CR25, from Avon. I used this on the Around the World in 80 Days Rally and many other long rough-road trips, and never had a puncture.

VOLVO GLT

This car morphed from the old belt-drive automatic Dafs, as Volvo bought up Daf and tried to produce a more conventionally engineered small car – the ultimate being the 2-litre Volvo engine shoehorned into a car that had previously used lighter Renault engines. It had a terrible press, condemned as the worst understeering car on the road, as most of the weight of the heavier engine was well forwards of the front axle line (Audi-style), and as soon as you tried hard, the front end just ran wide ... and then some. Yet there are at least two out there on the British club night-road-rally scene, and both the crews swear that a) this is strong, reliable and ultra-cheap rallying, b) the cars can cost around £200 if you look hard enough, c) the ultra-soft seats should be left alone, they are brill, and d) all it needs is stronger front springs, a sump shield, belts and a map light.

For build quality (not having the wiring loom fall around your ankles, as can happen with some cars (Sierra)), and nothing dropping off, rear-wheel-drive and a 2-litre gutsy engine, it's a car that is clearly underrated. You just need strong biceps and bravery pills to hook it round corners.

CITY ROVER

Condemned by *Top Gear* before it was out of second gear, the car was a final attempt to refill coffers in Rover dealer showrooms. It was made in

A City Rover going home ... London to Delhi for two Southampton students, trouble free.

SPEEDPRO SERIES

India, designed with a lot of help from Fiat (floorpan and suspension is very Fiat-looking), with a Peugeot-designed engine. Now a dead cheap secondhand buy, and few newish cars come as rugged as this. It sits higher off the ground than most.

Chris Cardwell and Nick Clarke of Southampton University's motor club drove one back to India, having had just 20 hours of preparation from Simon Ayris (Volvo GLT seats, no sumpguard, waterproofed electrics, Amsoil grease in the wheel bearings, Castrol Magnatec diesel oil in the engine and a decent pair of wiper blades were about the extent of it). The rubber sealing strips (in fact, anything made of rubber) are the worst aspect of the car, which is simple and rugged. It got to India without any troubles apart from having a dent tapped out of a wheel, and a sticky throttle cable. It ran on five-ply Continental Reinforced-type tyres.

MORRIS MINOR

The Minor still pops up on events like the Mongol Rally, which is perhaps a shame as the car is left at the end of its journey.

The ex-Archbishop of Canterbury's Morris Minor, which I took on the first Himalayan Rally in 1980 with the Rev Rupert Jones, had very little done to it. The aim was a finish. The spec was a 1275 Gold-Seal Sprite engine, ported and balanced, two SUs, an MGB oil cooler mounted well back on the inner wing (only to give it greater oil capacity), Allegro expansion bottle for the standard radiator, straight-cut gear set, standard axle and saloon car spec new rear springs, Spax telescopic damper kits front and rear (from Demon Tweaks), standard torsion bars, standard lever arms retained with Duckhams 20-50 engine oil, double-skinned sump pan plus steel sumpguard, flexible exhaust using rear box from a Rapier 1725, and

Russell Steel on his first ever rally. He finished 2nd in class on the Classic Marathon to Marrakesh in this Morris Minor, with Willy Cave on the maps.

the centre box of a Triumph 2.5. There was an additional fuel tank from a 5-gallon minivan mounted over the axle line at the front edge of the boot, with the standard tank retained. We used steel van wheels, 4.5J, 165-14 Avon Arctic Steels. Body strengthening was virtually nil – stronger sills from the convertible, and bash plates and caps over things like the rear spring hangers.

It got up to 6th overall at one point, and finished 10th overall, first in class. It probably never had more than 75bhp, but did well chiefly because all the other cars, such as the Opal Manta, Toyota Celica and the like, needed constant stops for servicing.

FORD ANGLIA

A lot can be done to the old Anglebox. A firm called Milton Racing at Ashford, Kent, has an Escort-style World Cup crossmember to support the engine, and much else. Ford Escort inner wings can be made to fit, which means the front strut hangs down at a much more conventional angle (less king-pin inclination, for the technically minded), and makes it easy to fit Escort discs. A Sierra Type-9 5-speed gearbox would then transform the car. However, this is a lot of work, and you would be advised to check with the Historic Rally Car Register how far you can go if you are using it on British events.

This is the HRCR listing of an ideal British club car spec: engine – 997,

Ford Anglia – lots of bits about, but take care not to turn it into an Escort if you want to do the HRCR Championship.

TWENTY BARGAIN PROJECT CARS

1200, or Cortina pre-cross flow 1500. Single or twin-chock carb. Gearbox – 4-speed, cast iron casing, no dog box gears. Steering box – no rack and pinion. Brakes – solid disc, two-pot cast iron callipers, drum rears, single piston. Adjustable brake pedal bias, not reachable by the driver when belted in. Suspension – non-adjustable platforms, narrow two-bolt mounting of the anti-roll bar, not wide four-bolt Escort spec. Rear leaf springs, lever arm and telescopics. Solid track control arms. Negative camber can be obtained by drilling the crossmember. Axle must be be three-quarter floating, not Quaiffe, alloy casing OK, limited slip diff OK, 6in rims, 70 profile tyres minimum, central fuel tank in boot ok. Fibreglass bonnet, bootlid, Perspex windows.

A lot of Ford items are interchangeable with other models. Check out www.oldfordautos.co.uk for a range of parts – things like heat-protected clutch cables should be on the shopping list if you are building a rally car.

FORD KA

Early Ford Kas came without power steering and were a joy to drive, largely thanks to the Mini-like principle of 'a wheel at each corner,' and the Fiesta-based chassis but with a shorter wheelbase.

However, the main problem with turning it into a rally car is that getting bits is not exactly low-cost. There are no uprated off-the-shelf springs and shock absorbers, a sumpguard could be quite pricey, and while polypropylene bushes for the front anti-roll bar are cheap to source and help sharpen response, transforming one into a club car is not easy.

The biggest weakness is the gearbox mounting – ordinary garden chain, looped over the end and around the chassis rail, has been used as a last

The Ford Ka is not easy to modify, but this example is working hard on the first Welsh Endurance Rally.

resort to stop too much rocking on ultra-soft engine and gearbox mountings.

A car with a short wheelbase always suffers on long-distance events, as it bucks and twists over corrugations. The Ka has enjoyed a long life almost unchanged, relying on a 1300cc engine that owes its origins to the Ford Anglia, first introduced in 1959. Given that parts to uprate the car are not plentiful or cheap, it's probably best avoided – however, they are strong and simple.

THE MINI

The Mini is not quite low-cost, as any old-school Mini is now an appreciating collector's item. In the same way that the 1950s belonged to the Triumph TR, the 1970s to the Escort, and the 1980s to the Peugeot 205, the 1960s belonged to the Mini ... every British club rally was dominated by hoards of them.

Past it? Not quite. In 2008 a Mini Cooper was still winning a major British club rally against modern cars, proving that when it comes to ultra-twisty and unpredictable lanes, the little brick can still lick the mustard.

There are two national magazines on the newsagent's bookstalls devoted to the Mini, and many thriving clubs. The London to Brighton for Minis only is probably the biggest single motoring event in the UK, as it attracts a line-up of several *thousand* entries, and there

Mini Cooper S with all the right bits ready for the start of the Lombard Rally.

63

are two very good suppliers in the useful contacts section at the back of this book – they know their stuff and can help you prepare a car for rallying. Brand-new bodyshells and panels are also being manufactured. On top of that, there are books devoted to tuning (try *How to Power Tune Minis on a Small Budget* by Des Hammill, also available from Veloce Publishing).

The downside is that the car never had much suspension movement, and suffers on rough roads – you have very little ground clearance.

Weller Wheels of West Bromwich makes steel wheels in both 10in and 12in sizes for £50. Traditional Minilite wheels made of magnesium are still available (but not low-cost). King Pin can kit you out with a tyre in the old Goodyear Ultra Grip pattern.

If going long-distance because you are a masochist, fit an A+ cast iron exhaust manifold – it's very efficient, and as the Allegro team in this book discovered, anything that runs cooler than a tubular manifold is a bonus. Under-body protection, and keeping it hanging together, requires some serious homework.

HILLMAN IMP

For: fantastic handling, enormous grip, terrific free-revving ultra-smooth engine, lovely pin-sharp steering, low-cost. Against: incredible noise, you have to be good to wring the best from it. The Historic Rally Car Register accepts Imps on its road or stage championships with: engine, 875 or 998cc; carburettors, Stromberg up to 1.75 or SU (but not late HIF type) or twin-choke Webers; gearbox, 4-speed synchro; brakes, drum (surprisingly good, full of feel), or disc conversion using Herald or Viva discs

Owen Turner and Graham Raeburn win outright on the Hughes Rally in Kent in this Hillman Imp.

(check out parts from Performance Braking of Monmouth, email info@performancebraking.com), rear drums.

SKODA ESTELLE

The rear-engined Skoda Estelle might be a rare sight now, but if you can find a good one, you are on your way to a fun-driving rally car. You can't knock the record – 20 class wins on the Lombard RAC Rally in 20 years, including several top 20 finishes, and on rough events, such as the Acropolis around Greece, the cars always shone. The team ran on Colway and Barum 165x13 cheap tyres, stayed in bed and breakfasts when all the other factory teams were in swanky hotels, and chalked up an astonishing reliability record. Forget the jokes – the wonderful full-of-feel agile steering and super traction make old Skodas fun to drive.

Here are some cheap mods: fit Volvo 140 rear shocks to the front, and Saab 900 Combi to the rear (TR7 units also fit the rears). Make a full length sump shield from a single strip of alloy, but mount it so you can detach it easily. Change the seats and the steering wheel, stitch a few extra welds to the seams in the front around the suspension area, and keep the front springs standard. If you can beef up the rears 10 per cent that would be good, and away you go.

You can lighten the car by ripping out lots of soundproofing – do it thoroughly and you can save a hundredweight, but this is not a suggestion for long events.

The Estelle engine went into the front-wheel-drive Favourit. It can easily be modded to produce over 80bhp at the wheels. A downdraught carb from a Ford Capri 3-litre Essex engine fits straight on, but needs a change of jets. The standard compression ratio is a touch high for long-distance events at 9.3:1, so if you are going to Mongolia, retard the ignition a touch and leave the head alone. Renault Five Turbo pistons fit, but take the bore out to 1360cc.

The Skoda Estelle does well in classic trials, and is cheap and easy to transform into a club rally car – it's different and it's rare.

Skoda Rapid passes the lake at Walters Arena on the 2004 Welsh Endurance Rally.

Chapter 15
Driving tips

You can't possibly learn how to drive your rally car from a book; this only comes from experience. However, driving on ice and fresh snow, either in the Alps or the Atlas Mountains, or driving the sands of the Sahara (and you don't have to go much further south east of Marrakesh to suddenly find long stretches of sand that have blown right across the road) – these are fresh challenges and experiences. Often as not, you get one chance to make a reasonable fist of it, or you could be there a very long time. Water crossings are also a daunting prospect if a wide river is suddenly facing you, and this is something you have not driven before.

So, here are a few survival pointers.

SAND

This is the hardest of the lot. It requires a deft, ultra smooth touch, and a great deal of mechanical sensitivity if you are to avoid getting totally bogged down (but

Getting stuck can be a lonely experience.

we have all done it – even the very best get stuck).

The front wheels are every bit as important as the back. If you veer an inch left or right and rub the inside of the tyre into a ridge of harder sand, you are dragging away precious momentum, and momentum is everything. You will pray that back in the workshop, the underfloor was made as smooth as possible, as anything that is now sticking down and running at right angles to the direction of travel, for example, a crossbeam in the chassis floor area, spring hangers, a not quite straight exhaust ... all these things can combine and gang up on you. You need an ultra-sensitive, smooth right foot on the throttle, and the ability to resist the temptation to suddenly snatch a gear. Keep up the revs, keep up the momentum – the moment you decide to change gear you are done for, as you break traction. Concentration on keeping the wheels dead straight is vital. The front tyres will act like dozer-blades the moment you don't keep things *totally* straight.

Sand varies. The darker patches that have taken longer to dry in the morning sun are crisp like digestive biscuits, and the car can get up a fair lick of speed and plane across these patches, which offer slightly more grip and less of a sinking feeling than the lighter ones. Read the surface!

Getting stuck and getting going again is a big test of determination. If

SPEEDPRO SERIES

A Mercedes struggles in the Sahara on the Around the World in 80 Days Rally.

Trust the locals ... the Hunter of Freddie Giles went on to win.

others are speeding past, you feel pretty deflated. Talking of deflation, letting down the tyres can help (but, have you anything to pump them back up with?), as you close up the tread pattern when you do this, presenting a nice flat bald surface to the sand, which is what you really need in the first place – the best sand tyres are totally bald ones. The last thing you need is any remotely chunky pattern, as this will only bite into the surface and rip down. What is more, deflating also widens the tread area by quite a significant percentage.

If bogged down, you need to dig some way ahead of the front tyres and clear the rear wheels, and ensure the driving wheels are not just going to dig down and make the whole situation a lot worse. This happens. Cars get stuck, the crew get out and think they are digging themselves out but it all comes to nothing the moment a driver gets back behind the wheel and gives it a load of welly.

Watch what the locals do. You never stop learning at this game, and on a trip to Mauritania during the World Cup Rally to Dakar I watched two locals get out of an old Peugeot 404 and look at a car that had rather made a mess of turning around on a perfectly good stretch of tarmac road, run wide and gone onto the verge, immediately sinking into soft sand. They didn't push and pull from in front and behind. They stood on either side of the car and rocked, crossways. Interesting technique this, as for one tiny split second, the wheels on each side were being rocked upwards just enough for sand on either side of the trench to fall under the tread of the tyres. We are talking a spoonful of sand here, being pushed under each wheel. Through constant rocking, the hole the tyre was in was gradually filling up, and the car was lifted. Once up a couple of inches, it was high enough to drive straight out.

Coconut matting or sea grass is good for interior mats as they are so useful – they grip into the sand whilst rubber mats just spin out. Steal the door mat before leaving home (the chewed one in front of the kennel is even better).

Never use wipers in desert conditions, nor the washers, as they just create an abrasive paste that will seriously scratch the screen.

A really strong tow rope is also a good idea, and so too having a bull-bar or similar to hook it onto at the front, something not buried under the bumper. The number of times we see cars that are set on crossing a desert, but ignore this fundamental, is quite astonishing – it's an error in workshop preparation that is repeated time and time again. When the standard tow-hook is buried in sand, you have only to dig that bit extra to find it. Hooking it all up and tying a decent knot in sand, in the full heat of the desert, only adds to the misery of this whole experience. A case here of Prior Planning Prevents Piss Poor Performance.

SNOW AND ICE

You need to be ultra, ultra smooth. You need to get the braking done when the

This sisal doormat cost £2.50, stops your trainers from becoming molten rubber on a hot floor, and is great to put under the wheels when you're stuck in sand.

DRIVING TIPS

car is travelling in a straight line. If you are still going too fast, get *off* the brakes. That's the hardest thing, as you have to overcome your instinct here. But you need your steering to get you round the corner and you risk locking up the front and just ploughing on straight across the road into the far bank if you do not overcome the instinct to keep pressing the middle pedal. If you have rear-wheel-drive you can balance the whole act by using the back to push the car around, and if you have front-wheel-drive, you have the advantage that powering out of a corner actually pulls the car around, literally. The only difficulty is that if you apply too much steering lock, you can push the car with enormous understeer. Come back with the steering, regain traction, and the car will begin to turn in.

The handbrake on a rear-wheel-drive car can be used with a little dodge that works in certain circumstances: snatching it off and on very rapidly when you are crawling uphill at walking pace on sheet ice can actually help you. The handbrake will instantly lock the wheel that is spinning the most, allowing the differential to transfer its drive to the wheel that is spinning the least. So, if you know you are on patchy, rutted ice, for example, you can use this handbrake technique to get the most power to the wheel that at that particular second is doing the most gripping. And inch by inch, you can move along. It's tricky to describe, but in certain circumstances it's surprisingly effective.

As with sand, smoothness is everything.

RIVER CROSSINGS

This is not part of the everyday driving experience. Too many drivers say, "ah, it's a rally car, charge in!" Wrong. Do you know if there are rocks under the surface? Do you know how deep it is? If you know cars have gone before you

Gently does it ... this Sebring Sprite descends an icy col on the Monte Carlo Challenge.

Cautious approach up the Todra Gorge in Morocco makes sense – Peugeots have their air-intake low down at the front – unless modified, an engine full of water is an easy mistake to make ...

(as on a Special Stage rally), splash on – providing you have taken to heart the notes in this book about waterproofing, and you are certain the fan is not going to turn into a motorboat propeller and pull itself onto the back of the radiator, you will gain a second or two here.

Caution on long-distance events, however, calls for a more prudent approach. The object is to get to the other side and not have to stop in mid-stream, and definitely not on the far bank (usually covered in mud dropped by previous cars). Similarly, you certainly don't want to be seen stopping a few yards further on, lifting the bonnet,

SPEEDPRO SERIES

Stalling the engine causes a vacuum in the exhaust, and sucks water into the engine – Mini on the Mongol Rally.

hunting for a rag and ... where the hell did you last put the tin of WD40?

So, less than walking pace going in, slipping the clutch, control all forward movement with the clutch in, out, a constant stream of revs, maybe around the 1500 mark, and go across as gently as you can while providing forward motion, with *no* bow wave going up the front of the car. A bow wave will only send water up to the distributor, and the higher the water in the engine bay the more the fan has to spray over the plug leads and coil – you might suddenly stop dead if the waterproofing is not up to much. Less haste, more speed, comes from getting it right.

At the far side you can now start to gun it. You want to storm up that bank, press the brakes a few times with the left foot to dry them out while the right foot keeps the revs up – water is now pouring out the bottom of the doors, the sills, the box sections.

If you have to be the good Samaritan and stop now to help a teammate, for goodness sake make sure you do so in a way that is not hindering the cars behind, because they don't have brakes to suddenly stop to avoid you.

Fans are best disabled if it's a long crossing. I've had water up onto the top of the bonnet of a Peugeot 504, and still the engine kept running on all four, despite the fan spraying water everywhere. We were lucky, frankly.

Splash plates around the side of the engine bay to stop a great whoosh of water riding up the side of the engine block will make a critical difference (a sumpshield also helps to prevent this).

Wind up the windows and put the heater on full – even a slight increase in cabin pressure may stop the water coming in. Finally, the jet of water from hitting a deep splash at speed is enough to wash out the grease from your wheelbearings – unless you have specified waterproof grease (Amsoil being the best you can get). Another good reason for prudence.

BRAKES ... BREAK IT
One of the worst things you can do is hit gulleys and sudden potholes with your foot on the brake pedal. Let your uprated suspension cope with it as best it can. Rattle over it too fast, and have everything on the back parcel shelf hitting the ceiling. Put up with sailing through the air and landing badly. But whatever you do, you must, must, must avoid sudden last minute braking, as that just sends the front of the car downwards, limiting the amount of suspension travel. Seriously hard braking when you have just seen a two-foot hole in the road will only add to your misery, as you now have no suspension travel to absorb what you are about to hit.

Stay off the brakes – let the suspension do its best to cope, and put up with your co-driver cursing you afterwards for not looking ahead and scrubbing the speed off ... they will only tell you that good drivers predict disasters like this. But last minute braking isn't going to win their praise either.

A deep water crossing catches out this Escort crew on the Bullnose Rally. A bow wave develops at anything over walking speed, then it's the navigator who has to

Chapter 16
Life on the road

LIVING WITH TRUCKERS

Long stretches of soft sand, or coming round a corner and finding the sun has not got to this bit, which is now suddenly sheet ice, or having to get across a river as wide as the Thames at Tower Bridge ... these hazards are nothing compared to the constant fight for your patch of road among the trucks that own it.

Old, poorly maintained trucks, heavily overloaded, pushing on to meet their own schedule are in a rally of their own. On long uphill sections, they will be thinking far ahead about where they might have to change down a gear, so will be wanting to build up speed on the approach. Car drivers rarely take account of the trucker's problems.On the other side, long downhill stretches become a nightmare for the truck driver, as brakes and gears are now combined to try to keep the thing from running away with itself. Overloaded trucks are like tankers at sea; they are not going to be able to swerve out of the way because you are going too slowly, or have stopped by the roadside. Trucks

Timing is everything when dodging trucks ... apart from the worry of what might fall off.

head straight at you and cannot give way, as the driver takes the view that

SPEEDPRO SERIES

Rickety wooden planks just about hold up a heavy truck – Outer Mongolia.

Trucking on in the wilds of Kazakhstan.

smaller lighter cars are agile enough to look after themselves.

Overtaking should be done briskly. Drop down into a lower gear and get the revs up before you pull out, not once you are alongside. Never, ever dawdle past – pull right out, as something might fall off the top of the truck onto your roof – this has happened. Don't expect a truck coming at you to slow down to let you cut in; he will maintain his speed come what may, so you have to plan all aspects of your overtaking manoeuvre. Be prepared to pull over suddenly onto the dirt shoulder to get out of the way – truckers will often force you to do this, but getting back onto the tar can be tricky. If you adopt a shallow angle of attack, the inside of the tyres will rub into the jagged edge of the tarmac and rip a hole in the sidewall of the tyre; you have to try and make an exaggerated turn so that the tread of the tyre takes the ridge more square on.

It's possible a truck moving slowly is belching out a big cloud of smoke, so you are like a Spitfire pilot diving into cloud and flying blind, certainly for a second or two if it's a dusty road – you cannot see the road ahead until you are almost past the truck. In moments like this, prior planning and thinking ahead is vital, as you can often 'read the road' and choose the moment to overtake, having decided that the particular stretch you are alongside will be free of potholes or ruts. In Africa, heavily overloaded trucks will wallow across to the 'wrong' side of the road on long open bends. If you have read the road and seen the way is clear, then nipping past here could be a sharp move, but is only done by planning ahead. If you are in a rally with the 1-litre rule, you really are exposed when it comes to overtaking. Building up speed, getting into the right cog and planning the attack is a must.

Tips like these are not going to be remembered in the heat of the moment – you need to develop an instinct for this sort of defensive driving. The most important thing is to keep out of other people's dust cloud (you are only clogging up your air-filter), and give all trucks a wide berth. Use whatever performance you have to execute an overtaking manoeuvre to full effect. Plan ahead on hills, do what the truckers do, and think about building up speed on the flat bit well before the start of the long grind.

It's all very different once the wheels stop, however. By all means take a break at the roadside truck stops, where you will find a friendly smile. Places where a bunch of trucks have stopped for roadside refreshment can be useful for your own comfort breaks, as you will find truck drivers know where to get the best omelette and a Coke. If you have a problem with the car, fellow travellers of the road will treat you with respect and encouragement. If you ever need help, such as roadside repairs, then a trucker is going to be your best friend, because he will know where and how to go about these.

MEALS ON WHEELS

So, despite treating yourself to the Cumberland sausage and mash in the restaurant of the P&O ferry, you are hungry already. Meal stops, and constant refuelling, will cut into your average speed – just as you think the car is eating into hundreds of kilometres, you see you have only covered an inch or two across the road atlas because someone in the car insists on stopping.

Is it possible to give the roadside cafes a miss and eat in for a change? It's a problem that has been on the agenda of drivers ever since they first tried to get to Monte Carlo inside a time schedule back in 1912 (the year Henry Royce became the first man to fly the Channel there and back without stopping – but, presumably, he packed a sausage sarnie).

Can you cook and eat as you drive? Tricky question. If you make a simple metal frame and have it welded to the bulkhead, so that a tin is less than an inch from the exhaust manifold, it's perfectly possible to heat up a tin of Heinz ravioli or baked beans in 20 minutes at rev-mark 3, or 4000 revs if you are in a hard-working A-series. The problem now is that on rough roads, the engine is rocking about on its mountings and is going to damage something if it starts hitting the cooker device. Are you making problems for yourself with something untested? A tricky question.

But it has been done, and on long-distance classic car events we have seen pre-war cars with just this sort of device warming up a can of water sufficiently for a cup of reviving Maxwell House. Tins of spaghetti are not light things to lug

LIFE ON THE ROAD

around, but the effect on an exhausted crew of something served piping hot is immeasurable. A French stick of bread goes stale in half a day inside a hot car, but wrapped in tinfoil and toasted with wedges of Brie it becomes a different thing altogether. Finding somewhere to wedge the bread – if that's all been thought out by the workshop beforehand as part of the preparation, a tired crew will be eternally grateful.

There are plug-in kettles that work from a cigar lighter, and they are great. You need a supply of plastic cups and spoons. In fact, a supply of paper plates is not adding to the weight and could come in handy from time to time.

Boil-in-the-bag emergency rations used to be gruel, but even army rations now have duck and orange, or a Thai green curry. You can find pasta dishes in tough bags to withstand yumping, available in kit bags at most good camping stores – not cheap, but worth the investment. Don't forget a simple camping stove, two or three throwaway lighters, and a box of matches. Mongolia is a high plateau – perhaps the oxygen is a bit thin, as getting a match to light isn't easy.

TOILET STOPS

Is it possible to take a pee while on the move? In case you are wondering "why hasn't someone thought of this before?" we ought to place on record the bold experiment for world science undertaken by Jim Gavin, in preparation for the 1968 London to Sydney Marathon, the first revival of long-distance rallying. In order to discover how to overcome the challenge, Jim and a friend decided on a very liquid lunch in Chelsea, followed by a drive down the A4 to Bristol for tea, in an open-topped Daimler Dart sports car. They were equipped with several black plastic bin liners.

The Slough to Reading stretch offered a nice straight section, and it was now time for the experiment. The only way Jim could comfortably exploit the bin liner to its maximum was to stand up on the back seat, behind the driver. Alas, even at a steady 40mph the breeze made holding onto a bin liner one-handed far from easy. A Mod, complete with fur-lined Parka, riding his Vespa Vibrator festooned with chrome mirrors, caught the black bag full of urine smack in the face, just as the Daimler darted past.

Jim Gavin tested a plastic bin liner from a friend's Daimler Dart while preparing for the '68 London to Sydney, as a means to avoid toilet stops – it didn't work ...

Gavin reports that the experiment was only partially a success, but it might be more successful from a Hillman Hunter – providing you still have a backseat.

INSURANCE

Most standard motor policies exclude racing or speed events in the small print, but thanks to the MSA, which governs the British club scene, there is a national insurance scheme in place that enables anyone who wishes to take part in a UK road event to upgrade their existing policy in order to obtain third party cover in a timed event.

There are specialists, such as Competition Car Insurance and Richard Egger, that could also sort you an individual policy, and if you have one of these, obviously you don't then have to sign up for the MSA-commissioned scheme.

Overseas events are different. If you are in an event with just a set route, with checkpoints that merely record proof of passage (passage controls), then all you need is normal tourist insurance, or what was known as a Green Card. A lot of insurance policies these days cover you for Europe. However, you need to check if they consider Serbia or Romania as part of their normal European cover. If not, it's not the end of the world ... you can normally buy insurance at borders, although it's yet another queue, and more hassle.

You should organise personal travel insurance that includes repatriation in the event of injury. A local travel agent could help you with this.

CALLING INTERNATIONAL RESCUE

Before leaving the subject of insurance, there is one insurance scheme that could

SPEEDPRO SERIES

Rear suspension failure, so hitching a lift – in 'Middle Earth.'

be of real help to anyone taking a car on a long-distance event – it's possible to take out breakdown cover via the RAC or AA travel departments, which includes return of your car – to your house. They in turn use specialist companies who cover Europe with a breakdown service, and you can buy this without even having to join the AA or RAC.

It will cost about £100, but it will cover your car to the far side of Turkey, which means it won't expire until you drive up the border of Iran. Break down terminally, with no chance of repair, as far away from home as this, and your car can be trucked back – good value!

MOBILE PHONES

Tell your service provider you are going away, in writing, otherwise you will be cut off. Texting is cheap – mere pence – so if you are taking a mobile phone, it might as well remain in working mode. Just about every town has a mast these days, even Timbuktu. If you ask your service provider it might actually hook you up to some countries not normally on your contract.

Credit card providers also need written instructions, as they might refuse your purchase of an emergency flight home from the middle of nowhere, after cancelling your card suddenly because they have decided some backpacker must have stolen it – confusion like this is best avoided with written instructions before departing, listing the countries you are visiting.

Paul and Sandra Merryweather enjoyed an easy win on the 2006 Classic Safari Challenge, due in no small part to the way their Mercedes ran with total reliability – here, £5-worth of wire rabbit-hutch mesh from a local hardware shop across the front of the grille and lights collects hundreds of butterflies. Easier to clean out than an overheated, clogged up radiator.

Note the Michelin Agila van tyre – longer lasting and more shock absorbing than Special Stage rally tyres, and half the price!

Chapter 17
Border crossings

A few words about the frustrations of crossing from one country into another. In Europe it's never been simpler, and this also goes for Estonia on the edge of Russia. Even Turkey is now a lot more simple and straightforward.

If you are entering certain countries you will have the car details written into your passport as a matter of routine, so you can't fly off home and leave the car behind. This might well be in addition to a Carnet de Passage; a booklet of counterfoils around A4 landscape size.

How does the Carnet system help you cross borders? It simply means you enter and exit different countries without having to pay import duty. Some countries might not ask for it, whilst others which handle tourists regularly will do, like Morocco and Tunisia. Many countries try to allow temporary import of vehicles with their own system – the way to check on what is needed is simply to call up the RAC travel department, which also

A bridge so far ... a Suzuki on the Mongol Rally enters southern Siberia.

issues the Carnets. It will want a bank guarantee that you are good for any claim made against it, and it will come after you hard if you leave a car behind that then triggers a claim for import duty. To satisfy it, have either a bank

SPEEDPRO SERIES

guarantee or take out an insurance policy.

It works like this:

You arrive at a border, going first through immigration where your passport is stamped "arrival" with a date. You move on to the next keyhole in a window, and customs will now ask for your Carnet. They stamp two counterfoils, tear one out for themselves, and hand you back your booklet with the other counterfoil containing the same stamp, now dated. Don't even think about moving to another desk if you are missing a stamp in the book – it will only cause problems later on. As you watch the process, ensure that both tear-out sections are each given the same "arrival" stamp. The same process then follows at the exit of the country, where they also keep one of the stamped counterfoils, handing you back your booklet with a stamp that confirms you have left the country. It's vital these counterfoils, a bit like cheque-book stubs, have not just a stamp to prove you were at these border posts, but that they are dated.

You might have to watch the vehicle be inspected. It's chiefly to look over the engine number and the chassis number, to ensure that they match what is written in the book.

The registration plate needs to be clear – if it's been knocked off at the last river crossing, you have a bit of explaining to do. Usually no big deal, as there is one on the back. Now the essential part of the inspection – is the chassis plate easy to find, does the number, or the VIN, show up clearly, and does it match the Carnet? You will need to know where the VIN is, so why not move the plate so it's easy to read, and likewise the engine number. This is usually a grubby little strip of tin right down near the dipstick, covered in grease.

An exhausted Minor in south west Sahara – a rally-prep exhaust can avoid hassles like this. Failure to cap the engine mounts is the most likely cause.

A Fiat nears journey's end, Outer Mongolia.

It's far better if you visit an engraver of the type who knocks up name-tags for dog collars, and ask them to make you your own engine plate, then glue it with Araldite to the top of the rocker box. If you don't have an engine number, or you have bought a replacement engine, you MUST get a plate that matches the vehicle registration documents (V5), as this is submitted when you apply for a Carnet.

PAPERWORK REQUIRED AT BORDERS

• Passport visas (if applicable).
• Carnet de Passage (if applicable).

BORDER CROSSINGS

A Suzuki on one of the few tracks in Inner Mongolia.

• Car papers – registration document (V5), driving licence, car insurance, UK road tax, MOT certificate.

DOS AND DON'TS AT BORDERS

• Don't be seen to jump queues.
• Don't be seen to take any photographs – all borders are no-camera zones, even though any spy could see anything he wants by satellite.
• Don't offer bribes. You only make it worse for those behind you.
• By all means offer to give your pen away – take lots with you.
• Be prepared to wait a long time – some border crossings can last all day.
• Fastest way to cross a border is go through the processing alone – do it with a bunch of mates and you go through at the rate of the slowest man. It only takes one person with a date in the wrong box or something that appears to be obsolete, and the whole bunch sit around while the argument goes on ... your chat-up and personal rapport will be better if you are alone.
• Don't ask for trouble! One crew on the Mongolia Rally checked in at a border and declared their occupations as beekeeper/philosopher. They were frog-marched off for two hours of questioning. Put down motor mechanic/tour operator, and you will be far more easily understood.

Hot work for Peugeot 205 crew on World Cup Rally to Marrakesh and back.

• Get some key prhases written down in Farsi (Arabic) or French if your sign language is not up to much, explaining what your aim is, what the event is, and where you are going next.

MORE USEFUL TIPS

Make a good colour photocopy of your passport and visas. Ensure the expiry date of your passport is at least 6 months after the planned end date of your trip. Leave a photocopy of your passport in a safe place at home. Take extra passport photos of yourself in case you need to apply for more visas.

Make a note of British embassy addresses and phone numbers. If you lose your passport a British embassy can help issue a new one far easier if they have something to go on. Replacement Carnets can't be done.

Obtain from the RAC an International Certificate for Motor Vehicles, a multilingual version of the V5.

An International Driving Permit (obtainable from the RAC and AA) is also translated into 10 different languages and includes a photo identity. This looks highly official, so will impress the local police if you are stopped on the road, and it avoids having to hand over your original passport or licence. This 'driving licence' has no legal validity, but is the most useful document you can carry. They cost around £5.

Take a good colour photocopy of the logbook and MOT certificate. Keep the original at home. A good copy (both sides) will be virtually indistinguishable from the real thing, certainly after being inspected at a few borders.

Conceal all documents, originals and copies, in two different places in the car. Place in a waterproof plastic wallet and hide in a money-safe under the seat, up under the dashboard, or in the roof lining.

A letter of reference from a bank may be useful to prove you will not be stranded through lack of funds. Also, take their phone number.

Keep a note of credit card numbers in case you need to report a theft, and carry some out-of-date credit cards to hand over if you are mugged.

Take some US dollars – they are acceptable currency everywhere. Keep the other cash hidden in a variety of secret compartments.

Chapter 18
Your health

TAKING CARE OF NUMBER ONE

A few words on personal preparations, aimed at those who are going more than a week in a car. If your goal is one of the four corners of the earth, then some preparations of a personal kind are as important as preparing the car.

Firstly, jabs. All doctors know what the World Health Organisation's suggested list of inoculations contains, and if you make it clear what help you are seeking when first making an appointment, you might find yourself being told what you need, rather than being asked what you think you want. Hepatitis jabs are a good minimum – look after your liver, that's a good bottom line.

Nothing is worse than the effects of Delhi belly, and as soon as one member of the team has a severe case of the runs, the car is going to stop. If you are in a hot country and you are constantly thirsty anyway, the worry now is that you are going to enter some pretty serious dehydration. If it's bad, the sort of thing you get over the counter of a chemist shop for a holiday in Spain is not going to do much – but it's better than nothing. What you need is a drug that is going to get you mobile again (and you want a response in hours, not days).

I have always found Lomotil to be very good, but don't be fobbed off with substitutes. Lomotil is a stopper, it freezes the gut, so needs administering with care, using minimal dosage. An understanding doctor will issue it, but it's often hard to get.

Dr Paul Rees, who has accompanied several long-distance rallies, brought onto the scene a drug he has used from his military experience, Cyproxin, which is as good as Lomotil but contains an antibiotic that fights the bugs. This has been brilliant. Once you have had a bout of Delhi belly, a packet of rehydration crystals that can be dropped into a drink of water will have you feeling vastly better (Dioralite).

Fortunately, despite some 20 trips to India and several to various parts of Africa, it's never hit me – but it has poleaxed colleagues. So, can it be avoided? Is this an inevitable downside of long-distance marathon driving?

Co-drivers tend to get it more than drivers. This might be because they do more shaking of hands meeting people or because they are constantly handling the local money.

Tubs of baby wet-wipes are a boon inside a car, or alcohol hand-rub, and providing you can get the lid back on and keep them sealed, you will be surprised at how long they stay moist. A regular wiping of the hands and dripping brow is a good idea, and if you run a wet-wipe over the steering wheel, door handles and gear lever fairly regularly, you might, just might, keep back local germs that your body is not ready to fight.

YOUR HEALTH

No shortage of drinks on hand for the workers – Plymouth Dakar crews help each other. Headgear is important.

If you don't keep a constant supply of water going through you, and don't get proper rest, you become 'run down' and then body resistance is weakened. Some flat Coke is supposed to be good at helping stomach complaints, and Schweppes Tonic Water is excellent (it contains a minute portion of quinine, which helps heal flesh wounds and is anti-malarial). Gin is also good.

Local bottles of water are sound to drink, but not if the cap is unsealed – that often happens in hotel rooms. If you unscrew the cap without breaking a seal, it clearly it could have been topped up under the tap when the room was made up. Carbonated tins or bottles are best. Stop where truck drivers stop – take a hot drink of mint tea (you can't go wrong with that, replenish your sugar levels). The toilets might be disgusting but you will find a phone that works, and drinks will be on supply.

Carry some tiny packets of salt of the kind you get in fast food restaurants like McDonalds, as a shortage of salt in the diet is one of the things that can make you really ill if the trip is days and days of being holed up in a car. And why not take the odd packet of mustard or ketchup? It might make local kebabs that much more palatable

Personal cleanliness matters. If you are camping, and aiming to visit a hotel every three or four days to wash out clothes, you are going to miss a long soak in a bath or a refreshing shower. On the subject of washing and clothes, pack a freezer bag (with zipper) of hand-wash soap powder, and hang clothes over the radiator or out of the hotel window for an hour or two each night. This way you don't need to burden the car with loads of clothing, which you certainly will wish to avoid having to lug up and down stairs of the Hotel Fleapit.

Outdoor clothing specialists, such as Rohan Clothing, produce cotton trousers and shirts that drip dry in three hours.

A good many rally crews on long-distance events complain of 'heat itches' and something akin to athletes foot between the toes (it's not athlete's foot, but the burning and constant itch is just as annoying). Hot water doesn't seem to have any effect. Actually, getting properly clean gets harder – the more you sweat and the more you endure long hot days cramped up inside the car, the less effect soap and water seem to have. This is in the main down to your choice of soap.

Pack a good first-aid kit, and if you are on an event which does not have a qualified doctor, do some research of your own and go on a first-aid course. You only have one body and one life, and events that are not backed up with fully qualified medical care and a get-you-home rescue plan are putting all of the responsibility onto your shoulders.

Before setting out do see a dentist, as fillings can rattle loose. On one event we watched as one driver had a tooth extracted with a pair of pliers from the mechanics tool kit.

Binoculars, sunglasses, money belt, sun hat, suntan lotion, secret compartment for copies of visas, passport details, universal plugs for sockets and sinks and a decent bar of soap (carbolic is best) should all be on the list of personal kit.

Finally, some advice for your own safety. Some events will draw a line on a map and encourage you to drive it, without any thought whatsoever about whether it is actually safe for outsiders to go into these remote areas. Regions of civil unrest, recent war zones, or abject poverty where people are almost starving, ought to be avoided. If the British embassy in Pakistan is saying that a rally crossing the country is not

SPEEDPRO SERIES

Cool dude: hats are important in the fight against heatstroke. Here, a snappily-dressed Alastair Caldwell changes a wheel on his 1300cc 205, on the World Cup Rally around Tunisia. He went on to win ...

Mobile satellite phones can be really useful in moments like this. The Flying Doctor Service was called up, and had to land in the road to take away an injured rally driver on the Classic Safari Challenge – the organisers had to set fire to a spare tyre to guide the pilot. The first on the scene was another competitor, who thankfully was trained in first aid.

advisable, and, in even stronger terms, suggests this is highly risky, it is basing an opinion on its own knowledge of being in the country. This advice is not given out because it wants to discourage travellers and avoid the problems of lost passports, no money, robbery or petty thieving – that's what embassies deal with all day long. If it says a border is difficult, closed, or, an area is really out of bounds to group travel, or high profile rallies that have advertised themselves to the rest of the world on websites, don't go. This advice has to be respected – it's not fun and not a challenge to just adopt an attitude of "let's ignore officialdom". Not for nothing has the vastly funded and super-organised Dakar Organisation moved right off the continent of Africa for South America.

If you are told "don't come" by the embassy or consulate staff and then choose to ignore it, consider this ... you are not going to get very far up their priority list when you get into trouble – and then have to call them to help you.

Is Immodium any good? Any serious drug sold freely over the counter cannot reasonably be considered to come in max-strength. However, the problem with drugs designed for too much rich food on a Spanish holiday is that the makers presume the sufferer has plenty of time to loll around. On long-distance rallies, a more immediate cure is demanded. Immodium is better than nothing, but the downside is that it often just bungs you up, takes days to work, and only puts off tackling the cause of the problem, which then returns when the effects of Immodium wear off. So don't be surprised if its effect is temporary.

A far better idea is to approach your doctor. If he won't give you Cyproxin or Lomitil, it's possibly because he thinks you can't be trusted – if he only fears you will overdose on it, ask him to limit you to just six. A few tablets of the proper cure is better than none at all, but getting hold of the best drug is often not easy ...

On the 2008 Cameroon Rally one crew went to Tunisia only to find that there is no access to Algeria – it's off-limits – so were turned back. They could have easily discovered this fact from the internet or called up the British embassy first.

Chapter 19
Sponsorship

A REALITY CHECK

Nobody is going to give you money for a sticker on a car. Every day there are World Champion wannabes knocking on doors hoping someone will pay for their hobby. Here are some down to earth pointers:

Give value for money. If you reckon you can go into the local newspaper, local radio, and just before the off, the regional TV, then plan the publicity campaign and go for the sponsorship – not the other way round. It's no good using a bit of local fame and then calling up would-be sponsors with "you might have seen me on the news last night," as you have just blown your one big chance of giving someone some value.

You could embrace local schools. Ask a local head if you can visit and put the car in the playground, and talk about the geography of the route – and plan a return to show how you got on and photos of Mongolian Yurts, or nomads in the Sahara. Get children around your maps. A sponsor might see this kind of 'community involvement' as an incentive to put their name on your car. A local supermarket might well like this subtle appeal to reach local families.

Write letters that are short and to the point. Call up to ensure you get the letter on the right desk, and name the person in the letter, not a general round-robin "Dear Sir, I am wondering if you might be interested" etc. That's wet, and you don't want to sound like a drip.

Don't be vague. Busy people want you to get to the point, quickly. They are interested in publicity benefits, not your chances of doing well – be prepared for answering tricky questions like "what if you break down on the first day?"

If you have any past track record, that is reassuring. Cultivate a relationship – sponsors get involved because of a personal interest. A company sponsoring a major golf tournament does so because a) the company can take guests to socialise and do business over lunch, and b) you can bet the company marketing director plays golf. Same for yachting – big bucks here, and all the photos look the same, angry seas and half a sail up, which may or may not have a logo visible. But the boss who signed the cheques goes boating with his kids, or probably takes sailing holidays.

Your task is to find someone with an interest in cars. If there is a charity angle, push that for all it's worth – be upfront and honest about what goes to the charity, and what is going into the car preparation. Sponsors understand that a better funded car is better prepared, and has more chance of success.

A local company might suddenly take an interest if it can see that you could produce a dozen photos of the livery on your car in unusual, far-flung places, and produce an end of year calendar – most local printers could turn your photos into a calendar, which a company then sends out to customers.

SPEEDPRO SERIES

This is just one example of how a mutual relationship can benefit both partners.

Landing a sponsor takes time, a thick skin, and a ton of determination. Nobody in the commercial world is interested in rusty junk with zero-image, so bear that in mind ... presentation is everything. But, if you are off to conquer the world, you can bet your local newspaper and regional TV will be very interested – get something on a reporter's desk several days before they go to press, don't leave it until the day before the paper appears – and put plenty of contact phone numbers at the bottom, with all you have to say on just one page.

Next, follow it up with a phone call, "did you see my press release?" You have to be pushy in life, others out there are pushing even harder than you. This is a competition, and the competition starts well before you reach the start line.

Great sponsorship comes from "someone I know, happens to know someone who ..." In other words, a personal contact. Once, a girl wanting to drive her first ever rally asked a company director over a dinner party table if he had "a spare £100,000". As it happened, the company was up for sale, and he was seeking a buyer. Next day she got confirmation there was £80,000, providing she could get into a Sunday newspaper colour supplement – she pulled that off, and landed the big cheque, which more than paid for the car and the rally. The company director benefited, as the racy image in the national press was seen as prestige, coming at just at the right time. Others have got fifty friends to each chip in £50, and put their signature on the roof ... that's £2500 for the cost of supplying a couple of marker pens. Again, it all comes down to "who do you know?"

Finally, if all your ambitions totally depend on you raising sponsorship – because you can't afford it – you could be in for a big disappointment. Look at it this way: if you were taking up angling, or netball, or trying to cross the Channel under a hang glider, would you be looking for someone else to pay for your pastime?

The two girls celebrate a class win in their factory-sponsored Daihatsu, after a drive to Athens that became the first rally to enter Albania.

A well-supported Dacia heads into the Sahara for Bamako.

Chapter 20
Event guide

BRITISH EVENTS

Daylight events are very difficult, as there is too much other traffic around these days, so they tend to be runs for one-make owners' clubs, scenic tours, or even fuel economy runs.

Night rallies of up to 170 miles are timed to the second, and are very competitive, a real adrenalin rush. Besides being able to drive quickly down the lanes, it is essential to have a good navigator with you who is able to plot the route on the move whilst calling out directions as to which way to go at the next junction. The first objective is to be on the right road at all times, and the second to arrive at the time controls at your due time, penalties being applied for being early as well as late

These events can be 'closed to club', which means you either join the organising motor club, or are a member of another invited club. Club membership fees are a few pounds, usually £10 or so. So, no competition licence is needed. For the next level up you need a competition licence, 'National B', which can be obtained from the Motor Sports Association (MSA), Britain's governing body, and this costs £36. Once you have a licence, you can go from event to event without having to be a member of the organising club.

The bottom run is what is known as a 12-car, and as the name suggests it is restricted to a dozen cars, where the amount of work required by the organisers in terms of route authorisations and permissions is much simpler. These are very sociable events of around 50 miles which often take place in the evenings ending up at a pub. This is where to learn the basics of map reading, navigation rally timing and working together as a team. Good results come from practice.

Then there are navigation, road, and Endurance rallies.

All events are based on the Ordnance Survey 1:50,000 scale maps, and timing is to the minute. On navigational events the emphasis is on plotting the route from instructions handed at intervals along the route (known as 'plot and bash'), whilst on road events the whole route is often issued at the start, known as 'pre-plot'. Both events may include regularity sections, timed to the second, where a set average speed has to be maintained and the location of the time controls is not disclosed in advance, or off-road special test, also timed to the second.

Any standard road car can enter and there are certain constraints as to engine size and type, turbochargers not allowed, and noise which are all set out in the MSA blue rule book.

All road events are timed at 30mph; however, organisers have to get a result, so resort to using tricky methods of defining the route, as well as choosing the most demanding roads where time can be lost by all but the very best and most experienced crews.

SPEEDPRO SERIES

With four extra spot lamps, this Chevette would not be able to compete on a British road rally. MSA rules limit extra lights to two. Advertising is also banned on UK road rallies.

Finally, there is Endurance, or Enduros, and these are events for modern hatchbacks with a maximum engine size of 1400cc, or 2-litre non-turbo diesels. There are restrictions on what can be modified, all aimed at keeping the cars standard and the costs down to a minimum. You have to install a single roll-over hoop and full harness seat belts, and the suspension can be uprated.

As in all road events, no advertising is allowed, no more than two spotlights can be added, cars must maintain their original interior trim, the battery earth lead must be marked yellow, and older cars with a carburettor must have an exposed throttle return spring.

In the last few years about 200 cars have been built to this Endurance formula, and there is a lively championship embracing about half a dozen local club events, with events ranging from the Lombard, the RAC Rally of old, over four days, to Oxford Motor Club's Bullnose, a simple one-day out offering great value for money – the entry fee is around £120. The competitive element comes from being timed to the second on off-road test sections, known as 'selectifs', through forests, around old airfields, farm tracks and the grounds of stately homes. Competition is fairly even, as there is a very limited choice of tyre, which also helps to keep costs down.

Every weekend of the year there is a wide range of road events throughout Britain, and you are best advised to check out the road rally section of the British Rally Forum (www.britishrally.co.uk), or contact the Motor Sports Association (www.msauk.org, telephone 01732 765000) for details of your nearest local motor club.

HISTORIC EVENTS

There is a huge range of different events, road and stage, for older classic cars, and a thriving national club, the Historic Rally Car Register (HRCR). Check out www.hrcr.co.uk.

There are road events, from easy jaunt to tough night events, and championships which include various classes for experts and novices and also age categories – up to 1968, 1974 and 1981 - so, take your pick. If you are entering a championship you have to comply with one common standard on eligibility, so you need to check what you can and cannot do to modify your car.

Modifications should be those that were available in period and not with the wisdom of hindsight, incorporating modern technology.

There is also the Classic Rally Association (www.classicrally.org.uk), which runs one big event in the UK, the Rally of the Tests, which aims to recreate the atmosphere of a 1950s RAC Rally with off-road driving tests, and it has a terrific atmosphere. It also organises the Classic Marathon, a week long event to many European destinations, which has now been run every year for over 20 years.

And if you have ever fancied driving from London to John O'Groats, virtually non-stop and not by the most direct route, you should check out the HERO organisation's "Le Jog" (www.hero.org.uk) which is always held at the end of the year – bad weather and long periods of darkness make it especially challenging.

The village centre of Modave in Belgium is closed down for a sprint session for the 20th Classic Marathon, a popular event for old British sports cars like this Morgan. MGs, Healeys, Triumphs, Jaguars and Astons mix in.

EVENT GUIDE

Minis can still win! An unexpectedly snowy 2006 Gremlin Rally.

Julian Broadhurst's Minor only has 27bhp at the rear wheels, yet beat Escorts, 911s, TR4s and a Mini Clubman to lead the HRCR Road Rally Championship.

You compete for a medal, and very, very few have ever won a gold.

Minor earthquake rocks championship

Julian Broadhurst has rallied Vauxhall Chevette HSRs, Lotus Sunbeams, Sierra Cosworths, Opel Mantas... "and rallied all the events I wanted to do, Isle of Mull, French and Belgian rallies and the like," and has also been a co-driver as well as driver. Now he is in Historic road rallying with a Morris Minor, with only 27bhp at the rear wheels, "and really loving it. We get down a few beers afterwards, the camaraderie is great." Didn't he do that with the big-boys toys? "Don't be silly, I once sat next to a driver who took to each event his own dietician!"

The Minor has beaten Porsche 911s, and as this book went to press, headed the HRCR Road Rally Championship, ahead of a Triumph TR4 and a hot Mini Clubman.

With Avril Banks alongside him he is doing events that are a mixture of regularity sections and off-road timed-to-the-second special tests, driving to each event and driving home afterwards. The car cost £1250 from a local librarian, and needed two front wings and some chassis welding which cost a further £300.

The engine is 1098cc and standard, as is the gearbox, no limited slip diff. The back has lowering blocks which has cured some of the famous axle-tramp, the engine is a single SU, fridge magnets hold the stopwatches to the dash (he doesn't believe in drilling holes), the brakes are standard 7in drums all round, and there is a Brantz trip mounted low down. The seats are kept standard, – again, Julian won't drill holes – so Avril slides around a lot (as does the car – results like these don't come without it being on maximum attack everywhere).

"When it's twisty, we do ok, not much keeps up, it's only up hills and long straights that ruin it."

"It's like going back to your youth and the days of the Escort 1100, where you dared not to lift off, planned the braking far ahead, and found every corner a grin. This is the same – we replace the front brake shoes after every event, the rears last three or four events, we run open-block tyres that are ok on tarmac as well as loose surfaces – what we don't do is spend money. The thrill of beating expensive cars with an old Minor is just adding to the enjoyment of it all."

Julian reckons he can get a few mods to double the power to over 50bhp at the wheels – with better shocks and an anti-roll bar all being on the menu ... when that happens, the world of rallying could be in for a shock. You wouldn't want to be with the bloke in the Porsche 911.

THE BEST OF THE BRITISH EVENTS

For modern cars

Four of the best road rallies from the South West, Wales, East Anglia and the Midlands are:

The Carpetbagger – organised by Bournemouth & District Car Club (www.bdcc.org.uk) and based in Axminster (Axminster carpets, equals Carpetbagger), this is a traditional road rally with a long tough night section through the narrow lanes of Somerset and Devon.

The Border 100 – organised by the Welsh Border CC (www.welshbordercarclub.co.uk), it takes in a full night of the best of the classic Welsh lanes.

The Preston Rally – organised by Chelmsford and District MC (www.chelmsfordmc.co.uk), this takes old and modern cars and runs through the night in East Anglia – it's named after a local garage, nothing to do with Preston. Terrific off-road sections around the

SPEEDPRO SERIES

The navigator turns on the traction control on the Saw and Weld Rally.

An Escort working hard to keep on time on the Carpetbagger Rally – one of the top must-do events along with the Preston, but not for the faint-hearted.

It takes good waterproofing and sound preparation to put up with a tough night.

Expect the unexpected – the Proton of Frank and Mark Evans on the Drystone.

edges of vast fenland cabbage fields and long "whites" (so named as they are white on the map). You need a strong car as the roads and tracks are rough. Not for the faint-hearted.

Drystone Rally – organised by Mid-Derbyshire MC (www.mid-derbyshiremc.co.uk), this is no-nonsense, plot and bash, night road rallying with a good selection of 'whites' in the Derbyshire Peak District.

Welsh road rallying

Some of the best road events in Britain are in Wales. There are over a dozen really top-notch events, organised as full-stress, competitive press-on rallies.

How to get going? How to get into the groove? There is one club running a 'newcomers only' event, the Rali Bro Preseli, held around February when the weather and darkness combine to ensure it's not going to be a breeze. The organising club, Teifi Vally Motor Club (www.teifivalleymotorclub.co.uk), runs a good informative website that explains the terminology for each section and how Welsh road rallying is run. For example, regularities are different in most Welsh events – they are timed to the second, but a clever choice of roads ensures there is no question of being early, you treat them as a press-on section, just about everyone arrives

EVENT GUIDE

late, the organisers use farm tracks and tricky Welsh lanes, with timing-points cunningly placed ... as the Teifi website says, "everybody arrives late" – how late decides the penalty you get. No crawling along with eyes glued to the instruments – this is code for 'put the hammer down'.

You usually get 120 miles or more. A lot of clubs are now incorporating off-road, timed-to-the-second tests, or 'selectifs'. The Twilight Rally in 2008 had a number of different venues, offering a fantastic 80 miles of off-road competitive sections.

There is nothing antiseptic or mollycoddling about the Welsh night road rally scene – it's still ultra-cheap, with a rawness that comes from facing corners just once, in the dark, and where the person alongside you contributes at least 50 per cent, or more, to the overall results. Without a decent mate on the maps and clocks, even above-average drivers struggle to keep up.

For Historic cars

The Rally of the Tests – organised by the Classic Rally Association (www.classicrally.org.uk), it revives the spirit of the 1950s with an authentic-style RAC Rally for the older car, with a mixture of tests and road rally sections, always held in November. There's a brilliant atmosphere, and you have to enter into the spirit of it and conform to a period dress code. It has won awards as the best-organised road event in Britain. No licence is required.

For Endurance cars

The Audi South West Endurance Rally – organised by the South Hams MC (www.shmc.co.uk) in Devon, it has a well deserved reputation for being well organised, with its mixture of timed to the second off-road tests and probably the smoothest forest sections in the UK.

A Ford Anglia hustles up a hillclimb on the Lands End Trial, an event now 100 years old.

For everybody

The Land's End Trial – organised by the Midland Motor Cycling Club, it caters for bikes and cars, old and new. You need something rugged and robust and able to climb up spectacular and classic off-road hills, like Blue Bills Mine and Beggar's Roost, but the sheer endurance and agony of it all is worth it for the ending, a spectacular drive across Devon and Cornwall. Over 400 cars enter this one, which has been going for over 100 years. You don't see publicity, as the club charges an ultra-low entry fee, and struggles to cope with demand. A totally unique experience, and no licence required.

FOREIGN EVENTS

The world is your oyster. There is now an array of events, literally to the four corners of the earth. Kathmandu, Timbuktu, Dakar (or, now it should be called Goa, as the Senegalese government try to discourage a load of unwanted old cars remaining in their country) of the Plymouth-Dakar rally, and what is generally regarded as the best of the bunch for budget bangers, the Mongol Rally.

Banger rallies

As the name suggests, these are for old clapped-out cars purchased for a few hundred quid with very little preparation. This makes it more of an adventure and a challenge to reach the destination, and those vehicles that do make it are then auctioned for charity or given away to the locals. There is usually no set route or time schedule – just get to the end as best you can using your own resources, and don't expect any hand-holding for yourself or your car from the organisers. Entry fees are cheap, a couple of hundred pounds, and entries for the best known events are at a premium – keep a watch on the websites so that you can apply online as soon as entries open. Your chances of securing an entry are increased the more outlandish the vehicle and the better your ability to raise money for charity, which invariably benefits those less well-off in the destination country. Not all events are run for charity, and some offer cash prizes.

SPEEDPRO SERIES

Mongol Rally

The Mongol Rally has grown enormously in a few years to become the biggest rally in Britain – around 250 cars set out from London's Hyde Park, plus another 100 from starting venues on the continent, to meet up at a castle outside Prague. There are then a choice of different routes to Ulaan Baatar.

Most go through Russia, where payments to police will be regular, and a small minority go the full-on adventure route through Turkey, Iran, Turkmenistan, Kyrgyzstan and Kazakhstan, with two days in Siberia, before crossing over the western border of Mongolia. After this there are a lot of days off-road, which can be rough and extremely wild, with no roads or tracks. GPS is strongly advised if you want to complete the full route without spending days going in the wrong direction.

Organised by the Adventurists, there is only one rule – that the car should be 1-litre (though there are exceptions – see below). Older cars generally are favoured, although there is nothing to stop you taking a more modern car (one with air-bags, for example), and if you have something you think is a novelty and contributes to the event's image and publicity (an ambulance, hearse, London taxi and ice-cream van have all been been accepted as wacky) then you are allowed to do it in something a little less frustrating that 1-litre. The small engine ruling is to maximise the challenge – the organiser's don't want this to be too easy, and it appeals to young people who otherwise will struggle with insurance. The entry fee is £500 (at the time of going to press), and you have to make a donation of £1000 to the official charity chosen by the organiser. The Adventurists have raised very considerable sums for smaller, less well known charities.

Starsky and Hutch, eat your heart out. A Ford Fiesta arrives in Mongolia.

The route is generally safe, ending up in a vast, stunningly wild and beautiful wilderness. Pack a tent, sleeping bag and camping-gas cooker, and take GPS, as there are literally no roads. You can lose days going round and round, a bit like a rubber boat in the middle of the Atlantic. If you can find your way easily, you will enjoy it all the more.

The organisers also run events in India, South America and North Africa.

The 'Ruta del Sol' event starts with the purchase of a VW Beetle in Ecuador, which you then drive across the Andes, through rainforests, and across the pampas to the finish in Brazil, where the car is sold on.

The Africa Rally is a drive to Cameroon at the 'wrong time' of the year, and is probably the toughest, perhaps most foolhardy thing you could launch yourself into, as it involves getting around the Sahara Desert at the hottest time of the year, when blinding sandstorms are daily occurrences, and arriving in Cameroon in the rainy season. It's awesome, though not for the faint-hearted, and would require a well-prepared car – and very careful personal planning as well. See the website (www.adventurists.co.uk) for more details.

Plymouth Dakar

The Plymouth-Dakar was the first banger rally. From a haphazard adventure first

Fiat takes on the world.

run in 2003, organiser Julian Nowill now runs a much slicker ship.

You get an impressive book packed with information, including tips and advice from past entrants, and this is to be taken seriously – you go through an old minefield, and while the time of the year (starting just after Christmas) is the most user-friendly weather window with sandstorms on the coast run less likely, the sheer magnitude of trying to get an old car round the edge of the Sahara is enormously ambitious. You also get copies of the Brandt and Lonely Planet travel guides, plus two nights in Spanish hotels to ensure you meet up with others. You don't get far without teaming up with a 'buddy' for mutual support.

There is no limit on engine size. You can basically take any car, but it must be left-hand-drive, as the route crosses Senegal (not going into Dakar) and straight out of the other side into the Gambia, finishing at Banjul, where the car is sold to local charities. One car is banned – the Gambian government won't allow in a Trabant two-stroke, and they only permit left-hand-drive cars.

There are different start dates, as the rally is split into smaller groups – this helps smooth things over at borders and campsites, and makes it manageable. Many of the stopovers are scruffy, down at heel hellholes and that is part of the appeal – you step into a totally different

EVENT GUIDE

A Hillman Hunter wins the first Plymouth Dakar, on the final beach run.

world. Some knowledge of French would be an advantage.

The organisers also run an event to Timbuktu. There is no left-hand-drive rule, but 4x4 would be beneficial – while you certainly will learn fast, it will take good and determined sand-driving techniques to romp it in a two-wheel-drive car, although it's not impossible. Timbuktu is fabled, but the town is being filled with sand blowing in from the Sahara, and there is not much to see or do when you get there – except wonder about the difficulty of getting home (there are no direct flights – in fact, there are almost no direct flights to anywhere. It's still one of the least accessible spots on earth). Check out www.plymouthdakar.co.uk.

The Budapest to Bamako is a take-off of the Paris-Dakar, but is ultra low-cost, and involves driving 5000 miles in 15 days through eight countries. There are two categories, touring and racing. The racing category sees entrants collect points for completing stages, and the one with the most points wins. In three years the event has grown to 400 competitors, and is now establishing a London start. Entry fees vary depending on when you enter, but can be as low as around £300, rising to three times that for late entrants – it is usually over-subscribed. The event caters for wide choice of vehicles. Check out www.budapestbamako.org.

There is a huge choice of events to almost every conceivable destination in Europe. Events usually take a week and cost a few hundred pounds.

StreetSafari.com – organisers of the Home2Rome, Czechwrecks, Staples2Naples, and Calais2Casablanca.

Charityrallies.org – organisers of the Leicester to Athens and London to Ulaan Baatar.

Try searching the web for 'Ramshackle Rally', 'Valhalla Run', 'Scally Rally', 'Crumball Rally', 'Sucata Run' or 'StudentBrakeAway' (a budget version of the notorious Gumball Rally).

Further afield there is ...

The Babe Rally – Big Apple to Big Easy. Pick up a car in New York and then drive 1500 miles in 4 days to New Orleans (www.baberally.com).

Raid de Himalaya – Start from Delhi and drive up into the foothills of the Himalayas, negotiating mile after mile of twisty, demanding roads about as far up the map as you can go before you reach China. This is a proper timed rally, with stage (open road) and regularity running together. Run by Vijay Parmeer of Shimla, it's the best-organised event in India (www.raid-de-himalaya.com).

If you are the owner of a kit car there are events to Italy, Spain and Morocco (www.guildofmotorendurance.co.uk) organised by Liege Cars founder, Peter Davis.

Even those with micro-cars and bubble cars, up to 600cc, are catered for on the Liege-Brescia-Liege Rally, driving an authentic route over daunting Alpine passes such as the Stelvio and Gavia.

Is it safe? Always check out the government's advice to travellers on the foreign office website, or feel free to call up an embassy – this travel advice is not meant to discourage you, it's an honest assessment of the dangers and likelyhood of you being robbed, arrested or kidnapped – if they say it's risky, then don't risk it. The world offers plenty of other places worth exploring.

Just because an organiser draws a line on a map doesn't mean it's safe, or even feasible to drive – events that don't support you along the way place an even greater responsibility on your own shoulders to check safety and viability beforehand.

Chapter 21
Useful contacts

ENGINE BITS

Imp parts
www.malcolmanderson.co.uk

K&N Filters
01925 636950
www.knfilters.com

Kent Cams
01303 248666
www.kentcams.com

Maynard Ltd (engine reconditioning)
01453 833185

Mini Spares
01707 607700
www.minispares.com

Mini Sport of Padiham
01282 778731
www.minisport.com

Piper Cams
01233 500400
www.pipercams.com

EXHAUSTS

Jetex
www.jetex.co.uk

GOOD OIL

Amsoil
020 8737 0649
www.performanceoilsltd.co.uk

Millers
www.millers.net

SUSPENSION

Brost Forge Ltd
0207 607 2311
www.rhinoray.co.uk

Coil Springs of Sheffield
0114 275 8573
www.coilsprings.co.uk

Gaz dampers
www.gaz-shocks.co.uk

Koni
0161 355 1275
www.koni-shock-absorbers.co.uk

Leda Suspension
01335 348880
www.leda.com

Milton Racing (Ford Anglias)
01233 730959
www.miltonrace.co.uk

Powerflex bushes
01895 460033
www.powerflex.co.uk

Rally Design
01785 521871
www.rallydesign.co.uk

Spax
0869 244771
www.spax.co.uk

Superflex bushes

USEFUL CONTACTS

www.superflex.co.uk

www.autofive.co.uk (Peugeot 205)

www.europerformance.co.uk (Vauxhaull Chevette)

www.oldfordautos.co.uk (Ford parts)

www.performancebraking.com (Hillman Imps)

WHEELS
Weller Wheels (steel wheels)
0121 556 2910
www.wellerwheels.co.uk

TRANSMISSION
Reco-Prop (UK) Ltd, Luton (propshafts)
01582 412110

SEATBELTS
Luke
01424 854499
www.lukeracing.co.uk

SEATS
Corbeau
01424 854499
www.corbeau.com

SPARES
Car Parts Direct
www.carparts-direct.co.uk

Speedy Spares
01273 412764
www.speedyspares.co.uk

BRAKES
EBC
www.ebcbrakes.com

Mintex
www.mintex.co.uk

TYRES
Avon
www.cooper-avon.com

King Pin (remoulds)
www.kingpin-tyres.com

Sportway
01664 560444
www.sportwaytyres.com

RALLY CAR PREPARERS
AB Motorsport (Peugeot)
01926 421111
www.abmotorsport.co.uk

Andy Actman, Peter Rix
Lenham Sportscars
01622 859570

Brown and Gammons (all MGs)
High St, Baldock, Herts
01462 490049

David Johnson (Proton, Perodua)
www.bridgemotorsports.co.uk

Owen and Jamie Turner
St. Albans
01727 834004

Paul Jackson
Didcot
07747 620061

Peter Banham
Derby
01773 850024

Simon Ayris, Andy Inskip
Rally Preparation Services, Witney, Oxon
www.simonsrallyprep.co.uk

STICKERS & NUMBERS
Coba Graphics
01420 511255

Trident Racing Supplies
www.tridentracing.co.uk

MAIL ORDER PARTS AND ACESSORIES
Demon Tweeks
01978 664466
www.demon-tweaks.co.uk

Holden, www.holden.co.uk

TRAVEL ADVICE
Foreign and commonwealth office
www.fco.gov.uk

RAC Travel Dept (Carnets, international driving licences)
01454 208000
RAC breakdown recovery in 47 countries
www.rac.co.uk

NAVIGATION
Roamerlite Poti
www.thebasicroamer.co.uk

www.donbarrow.co.uk (equipment)

CLOTHING & CAMPING
www.penroseoutdoors.co.uk

www.rohan.co.uk

MAP SUPPLIES
Stanfords
Covent Garden
www.stanfords.co.uk

The Map Shop
Upton on Severn
www.themapshop.co.uk

BOOKS
How To Win A Road Rally by Alan Smith – available from Demon Tweeks

Peking to Paris by Philip Young – Veloce Publishing
www.veloce.co.uk

SPEEDPRO SERIES

Sahara Overland by Chris Scott – good routes, and bandit areas to avoid
www.sahara-overland.com

RALLY ORGANISERS

Association of Classic Trials Clubs
www.actc.org.uk

Classic Rally Association (CRA)
01633 263366
www.classicrally.org.uk

Endurance Rally Association (ERA)
01235 831221
www.endurorally.com
www.pekingparis.com

HERO
www.hero.org.uk

Historic Rally Car Register (HRCR)
01332 672533
www.hrcr.co.uk

The Adventurists
www.adventurists.com

www.mongolrally.com

www.budapestbamako.org

www.charityrallies.com

www.plymouthdakar.co.uk

www.streetsafari.com

UK ENDURANCE RALLIES

Audi South West
South Hams Motor Club
www.shmc.co.uk

East Anglia Mid-Summer
Chelmsford Motor Club (also organisers of the Preston)
www.chelmsfordmc.co.uk

The Bullnose Rally
Oxford Motor Club
www.oxfordmotorclub.co.uk

The Lombard Rally
Wales Endurance Rally Association
High Street, Llandovery
www.rallyhq.co.uk

GOVERNING BODIES

The Motor Sports Association (licences, regulations, how to contact local clubs)
01753 765000
www.msauk.org

INTERNET FORUMS

www.britishrally.co.uk

Now is not the time to have the engine conk out just because the preparation didn't include sound waterproofing ... the crew could be about to miss the ferry – just a typical day for this Fiesta crew on the Africa Rally from London to Cameroon, organised by the Adventurists. A Fiesta was first to the finish line – check out www.battlestarafrica.com.

Index

Air vents 14
Africa Rally 90
Antifreeze 27
Austin 1800 59
Austin Allegro 9-15
Austin Maestro 10
Ayris, Simon 11

Bodywork 16, 25-28
Borders 73, 74
Brakes 35, 36

Carburettors 23, 24
Cooling 16, 25-28

Electrics 16, 29-31
Engines 18, 19
Events 81
Exhausts 12, 19, 20

Fault-finding 55

Fiats 56
Ford Anglia 62, 74
Ford Ka 63

Gavin, Jim 43

Health advice 76, 77
Hillman Hunter 59
Hillman Imp 64

Insurance 71
Interior trim 16

Job list 15

Mercedes 60
Mini 63, 83
Mongol Rally 9, 86
Morris Minor 23, 62, 74, 83
Motor clubs 89

Navigation gear 51-53
Nissan Micra 56

Octane boosters 22
Oil options 23

Perodua 58
Petrol filters 21
Petrol tanks 40, 41
Peugeot 205 57, 75
Peugeot 504 61
Plymouth Dakar 86
Preparation experts 89
Problem solving 55
Protection 45-47
Proton 58, 84

Radiators 25-28
Rover 25 57
Rover City 61

SPEEDPRO SERIES

Safety equipment 54
Sand driving 65
Seats 41
Sill strengthening 11
Skoda 48, 64
Snow driving 67, 68
Spares list 17
Spill, Dominic and Vicky 10, 11
Sponsors 79
Stickers 48, 60

Strut-strengthening 43
Sumpguards 45-47
Suspension 12, 16, 32-34
Suzuki 38, 59

Tankguards 47
Thermostats 28
Tripmeters 53
Trucks 69, 70

Tyres 13, 37-39

Useful suppliers 88

Vauxhalls 60, 82
Vents 14

Wheels 13, 37-39

Competitors gather round to use the bonnet of a Jaguar as a desk, writing down a route amendment on the Classic Safari Challenge. A bridge on the road ahead has been washed away, and a course car has phoned marshals at a previous time control to give out a change of route. Moments like these are apprehensive for first-timers, as now they are forced to rely on maps, and run outside the comfort zone of the Endurance Rally Association's route book. But crews who have tackled a British road rally in the past take moments like this in their stride, which just goes to show that any kind of practice and experience can always be a useful benefit in long-distance rallying.

Also from Veloce Publishing –

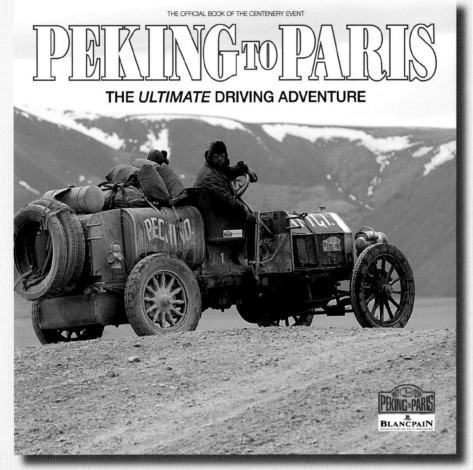

Hardback • 224 pages • Over 380 mostly colour photos
£29.99
ISBN: 978-1-84584-120-1

To mark the 100th anniversary of the original 'Great Race,' over 100 cars set out to drive the original route used by Prince Borghese in 1907. They ranged from authentic veteran Italas and vintage Bentleys to classic Aston Martins, and pretty much everything in-between. Drivers of 26 different nationalities came together to test their wits and their cars by driving for 40 days from the Great Wall of China, across the Gobi Desert and the wilds of Mongolia, through Russia and Eastern Europe and on to the big finish in Paris. Written by Philip Young, the event organiser, and featuring around 250 photos and official maps, this is a fascinating and colourful read for every red-blooded motoring enthusiast.

Prices subject to change • p&p extra • for more details call +44 (0) 1305 260068, email info@veloce.co.uk, or visit us on the web at www.veloce.co.uk

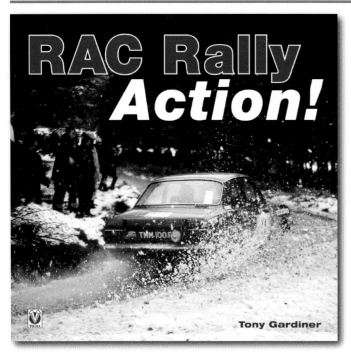

Hardback • 208 pages • 330 colour pictures
£35.99
ISBN: 978-1-903706-97-8

Here is an incredibly involving evocation of three decades of great motorsport. All aspects of the event are covered, including rare photos from manufacturers archives, hundreds of previously unpublished photographs, rally documents (regulations, programmes, road books, crew notes), and a full colour cutaway illustration of a famous winning car. This publication has full approval of the Royal Automobile Club and the Motor Sports Association.

Hardback • 96 pages • 90 colour and b&w photos
£12.99
ISBN: 978-1-904788-95-9

The predecessor of today's international rally sport, the alpine trials and rallies of 1910 onwards were an incredible test of endurance for early pioneers and their cars. Becoming ever more international, the event would continue in various forms until 1973. This book, written by a seven time Alpine Rally competitor, is an in-depth history of this incredibly demanding event. Illustrated with unique photographs and images of rally medals and trophies, this is an excellent book for the rallying enthusiast.

Prices subject to change • p&p extra • for more details call +44 (0) 1305 260068, email info@veloce.co.uk, or visit us on the web at www.veloce.co.uk

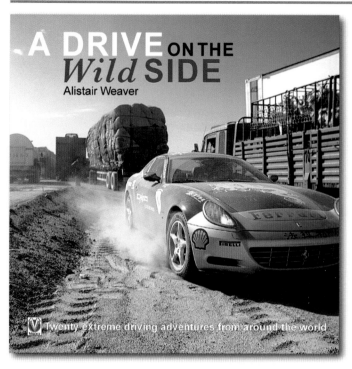

Hardback • 176 pages • Over 400 colour and b&w photos
£29.99
ISBN: 978-1-904788-73-7

Written by award-winning journalist and television presenter, Alistair Weaver, and illustrated by some of the world's leading automotive photographers, A Drive on the Wild Side, takes you on an extraordinary journey along some of the world's most challenging roads. This book recounts the fascinating, hair-raising and moving stories experienced during a career of automotive adventure, helped by 400 stunning photos. This book will appeal to car and travel enthusiasts across the world.

Paperback • 128 pages • 100 colour and b&w photos
£14.99
ISBN: 978-1-845841-28-7

The rallying Big Healey went down as one of the greatest road-rally cars of all time, winning the Liege with Pat Moss in 1960, the first all-female international rally, and the final event in '64, when the route went to Sofia in Bulgaria and back again with 96 hours of non-stop driving over the worst goat-tracks of Europe. The car also won the Alpine Rally. This book, with inside information, charts the making of the last true British sports car to win major rallies.

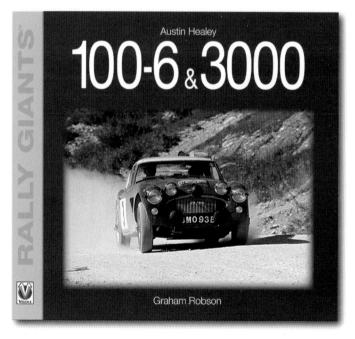

Prices subject to change • p&p extra • for more details call +44 (0) 1305 260068, email info@veloce.co.uk, or visit us on the web at www.veloce.co.uk

The Essential Buyer's Guide™

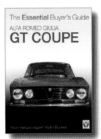   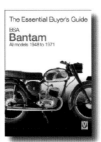  

978-1-904788-69-0 | 978-1-904788-98-0 | 978-1-84584-135-5 | 978-1-84584-165-2 | 978-1-84584-136-2 | 978-1-845840-99-0

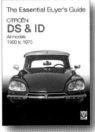     

978-1-84584-138-6 | 978-1-84584-147-8 | 978-1-904788-85-0 | 978-1845840-77-8 | 978-1-84584-200-0 | 978-1-84584-161-4

978-1-845841-19-5 | 978-1-845840-26-6 | 978-1-845841-13-3 | 978-1-845841-07-2 | 978-1-845840-29-7 | 978-1-845841-01-0

978-1-904788-70-6 | 978-1-84584-146-1 | 978-1-84584-163-8 | 978-1-84584-134-8 | 978-1-84584-204-8 | 978-1-904788-72-0

978-1-845840-22-8 | 978-1-84584-188-1 | 978-1-84584-192-8

£9.99*/$19.95*

*prices subject to change • p&p extra •
for more details visit www.veloce.co.uk or
email info@veloce.co.uk